SUNDIALS

MARK LENNOX-BOYD

SUNDIALS

HISTORY, ART, PEOPLE, SCIENCE

FRANCES LINCOLN

For Arabella

Frances Lincoln Limited
4 Torriano Mews
Torriano Avenue
London NW5 2RZ
www.franceslincoln.com

Sundials
Copyright © Frances Lincoln Limited 2006
Text copyright © Mark Lennox-Boyd 2006
Illustrations copyright © Mark Lennox-Boyd,
except where noted on page 140.
Photographic acknowledgments are listed on page 140.

A catalogue record for this book is available from the
British Library.

First Frances Lincoln edition 2005

ISBN 13: 978-0-7112-2494-0
ISBN 10: 0-7112-2494-3

Printed and bound in China
by Kwong Fat Offset Printing Co. Ltd

9 8 7 6 5 4 3 2 1

gnōmōn One who knows.
CLASSICAL GREEK

gnomon n. A pillar, rod, etc., which by its
shadow indicates the time of day; esp. the pin
or triangular plate used for this purpose in an
ordinary sundial.
SHORTER OXFORD ENGLISH DICTIONARY 1959

gnomoniste n.m. Celui qui s'occupe de l'art
de construire des gnomons.
LAROUSSE UNIVERSEL 1922

FRONTISPIECE Detail of the Holker dial, 1992, Holker Hall,
Cumbria, designed by Mark Lennox-Boyd.

RIGHT In 1965 the English sculptor Henry Moore was
commissioned by the owner of *The Times* to design a
sundial for the forecourt of the newspaper whose London
premises were then located in Printing House Square,
overlooking the Thames to the south. The area has been
redeveloped and the original sculpture was sold to IBM,
who have placed it outside their European headquarters
in Brussels. The artist retained an identical copy and on
his death it was purchased by the Adler Planetarium in
Chicago. This beautiful object, a perfect example of the
equatorial dial, decorates the lakefront and is ideally set
against its background of the most modern architecture.

INTRODUCTION

Anyone who reads this book may be surprised to learn that Isaac Newton spent more of his life writing about alchemy than mathematics, or that President Jefferson designed a sundial. Often people have interests that seem unusual and that play no part in their working lives. So it is with me, and I often ask myself why I am interested in sundials. When I was at school I was interested in two essential subjects for the diallist: maths and design, and in particular those features of design that are related to numbers. I remember writing a piece at that time about the double cube room at Wilton House, the magnificent English interior designed by Inigo Jones, and the relationship of the two lengths that were first known to the ancient Greeks and which are still called the Golden Mean, a proportion which has been used by architects since classical times as an example of two lengths in perfect visual balance.

Then one day in about 1965 I walked into a bookshop in Oxford and alighted upon the book on sundials by Frank Cousins, still one of the best and mentioned prominently in the bibliography. Today it sits covered with my scribbles, its cover broken. The acquisition of this book led to making my first dial, a slate slab, which still stands in my brother's garden. Since then, I have had many years to think and read about the subject at odd moments, even though my working life has had nothing to do with science, art or design. Latterly I have increased the proportion of my life I spend on the subject, as a result of this book and the considerable time and physical demands made by my project at La Meridiana, the large sundial I have designed and built in Italy as the stair tower of a house for my family, and which is described later in the book.

Perhaps this qualifies me to write for people who know nothing about the subject, and fear that they will find it difficult to understand. There is nothing really difficult in the book, and the mathematics are confined to an appendix. Nearly all modern dials are derived from designs that were settled hundreds, sometimes thousands of years ago.

The greatest sundial of all is the earth itself. The story of the sundial is the story of mankind's understanding of the movements of the earth around the sun. A sundial indicates the apparent movement of the sun in the sky, its position above us varying according to the time of day and according to date. Many sundials therefore tell date as well as time. Some indeed have been made expressly to measure date. The story is told chronologically, and contemporary designs are illustrated at appropriate moments in the text. Sometimes I have had to use unfamiliar terms, and I have therefore included a glossary for reference. The term 'dial' may be misleading to some people because it suggests an object that is circular, and most people think of the circular dials they have seen in gardens or on buildings. This is in fact erroneous, for the word is derived from the Latin *dies*, meaning 'day', and does not originally imply something round. Throughout their history, most sundials were not round, and the majority were not fixed in open settings. Only one conventional garden dial is illustrated in this book. This is intentional. Of the many hundreds of thousands made since the Renaissance, only a few have been designed for gardens.

It has been difficult to decide what to include. I have tried to describe most of the elements that are a necessary part of the history and science of the sundial from its origins to today, and illustrate those features with stories, anecdotes and pictures that are unusual, or which happen to be of personal interest to me. Books on the subject have so far mostly been written to explain how to make dials, or to describe the outdoor examples, or to illustrate portable or pocket ones with fine engraving. Curiously, few books deal with the subject comprehensively, historically, artistically or internationally, and none makes great mention of the personalities and lives of the patrons and makers – stories which provide striking examples of eccentricity, brilliance, charm, decency, megalomania, tragedy and even violence. This book is not fully comprehensive, but it does go wide, and, in my attempt to bring all the strands I have mentioned together under one cover, I hope to appeal to a larger audience than those who have written on the subject before.

BACKGROUND IMAGE Design for a sundial by Albrecht Dürer, from *Vnderweysung der messung* (Nuremberg, 1525). The title translates as *Teaching of Measuring*. This book introduced and explained perspective drawing for artists. This engraving explains how an equatorial, horizontal and vertical dial can be drawn with only compass and ruler.

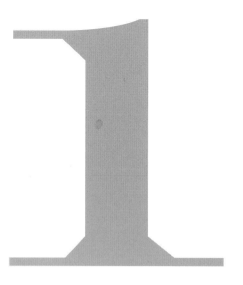

PREHISTORY

VITA IN MOTU
Life is in motion

No one knows when the sun was first measured, but prehistoric man would have been aware of its importance from the moment his brain had evolved sufficiently for him to understand that it caused the difference between warmth and the cold.

There would have been people of intellect among our Stone Age ancestors, individuals who, in their own way, were the geniuses of their time, endowed with superior powers of observation and deduction, which were enhanced over millennia as mental skills evolved and became more sophisticated. Man would have noticed at an early stage that the shadow of a vertical stick stuck in the ground varied during the day. Just after sunrise it was long, at midday short, and again just before sunset it was the same length it had been at dawn, but pointing in an opposite direction. He would also have noticed that the length of the shadow at midday was shorter at those times of the year when the temperature was hotter, and longer as the days grew colder.

Our imaginary early scientist lacked writing, instruments and calculation, but he had two advantages over modern man – he spent much more of his time in the open air seeing and thinking about his surroundings, and he was not in a hurry. Repeating observations many times before reaching conclusions of certainty was as essential to the primitive scientist as it is today. He and his successors would have noticed over many hundreds of thousands of years that the sun always reached its culmination – the highest point it attains in the sky – directly above the same feature on the landscape. This never varied. *Homo erectus* first appeared about 1.8 million years ago. Our ancestors had plenty of time to think about these things.

At night, he would have noticed that the stars appeared to revolve round one star – the Pole Star. At some stage, as his language developed, he would have invented words to denote the direction of the culminating sun – 'south' – and the direction of the Pole Star – 'north'. He would have noticed that the sun rose over different points on mountains or hills at different times of the year, and that as the days got warmer the sun rose and set progressively further to the north, until it reversed itself and the days began to get cooler.

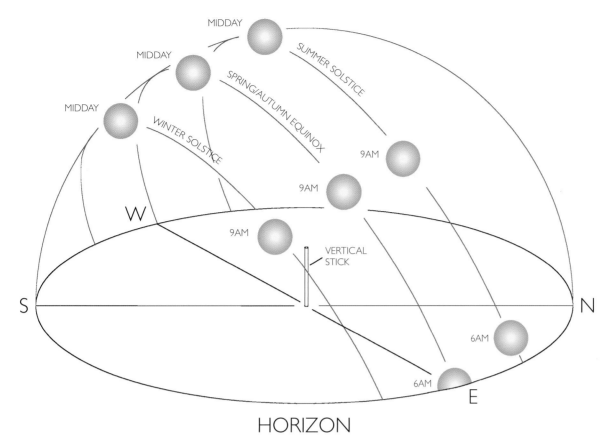

MIDDAY

MIDDAY

MIDDAY

SUMMER SOLSTICE

SPRING/AUTUMN EQUINOX

WINTER SOLSTICE

9AM

9AM

9AM

VERTICAL
STICK

W

S

N

6AM

6AM

E

HORIZON

South and north appeared to be diametrically opposite points on the horizon. If you join north and south, west and east will be at right angles to this line and you have a perfect cross with the four cardinal points of the compass. By observing that the sun traversed a longer sky curve in summer than winter he would have been able to reason that summer days were longer than winter days, even if the 'length of time' was a puzzling, indeed abstract, concept against which he had no yardstick for measurement. There came a stage when he reasoned that the length of time from sunrise to the middle of the day was as long as from midday to sunset. He also became aware of the equinoxes. The equinoxes are the two times of the year when day and night are equal in length, when the sun rises precisely due east and sets due west everywhere on the planet. If he lived in reasonably flat country, he would, by observation and reasoning, have noted the two points on a level horizon where the sun rises and sets on these occasions, when the days are not getting longer or shorter than the nights, but are of the same length, and that these points corresponded with east and west. This deduction would have been a major breakthrough in early scientific observation. There is no evidence for these assertions. Nothing is recorded from prehistory to confirm them, but the conjecture is reasonable.

While we learn our cardinal points at school, primitive man learned them in the wild. We are not normally up at dawn, and are hardly aware of the day of an equinox, but he would have become acutely aware of each equinox, for it signalled a change of season. The spring equinox means summer is coming, which is good, and the autumn equinox foretells winter, which is bad for people who are almost totally preoccupied like other animals with keeping warm and gathering food. The quest for food in a hunter-gatherer society requires travel, which in turn needs navigation for longer expeditions. An awareness of the movements of the sun, moon, planets and stars, however crude and imperfect, is essential for navigation and therefore of greater importance to the breadwinners of a simple society than a more advanced one.

Early man would have been aware of these movements and developed his thinking for many

The shadow of a stick stuck upright in the ground will, at the moment of the day when the sun is at its highest, point north; at the equinoxes it will point west at dawn and east at sunset. The sun rises and sets progressively further to the north until the summer solstice. Thereafter it rises and sets further to the south until the winter solstice.

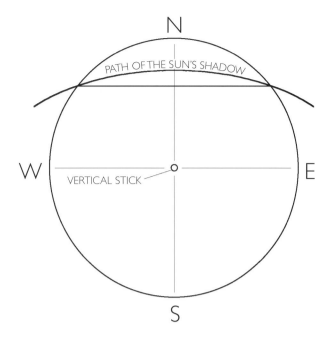

years before knowing for certain that the movements of the sun always held true. When he had certainty, the fruits of his observations and knowledge were passed on from one generation to the next. The mystery of it all must have been very disturbing indeed, which is one reason why in several ancient societies man worshipped the sun.

Different early civilizations observed and attempted to calculate the movements of the sun. The summer solstice at midsummer, the winter solstice at midwinter: these are the instants in a year when the midday sun is at its highest or lowest. Chinese legend has it that in their antiquity astronomers used a gnomon to measure the length of the shadow of the sun at the time of the solstices. These skills were developed independently in different communities. These activities must have been induced in part by the desire to know the season for agricultural purposes but one cannot discount mankind's abiding interest in understanding the mystery of the sun for reasons of sheer curiosity, heightened by the fear that one day this giver of life might not reappear. After all, it would have disappeared at least partially for some minutes at the time of an eclipse by the moon, a truly terrifying moment, the memory of which would have been passed from parent to child for many generations.

Neolithic and Bronze Age rondels (or circles) dating from as early as 4500 BC have been examined in the Czech Republic. These primitive earthworks consist of ditches and palisades, and several are so accurately aligned with the position on the horizon of sunrise on the days of the summer and winter solstices and the equinoxes that one cannot but conclude that these places were temples of the sun. Some of these earthworks, more than 300 feet in diameter, are shaped like squashed eggs when viewed from above. This is precisely the shape of the rising or setting sun, which appears flattened by atmospheric refraction. It has been suggested that the rondels were designed to look like the setting sun and to be a message from humble man to the divine sun, giver of life.

There are several large stone circles to be found in the United States. The Big Horn Medicine Wheel in Wyoming was originally constructed by Native Americans in 1500 BC and was used to predict seasonal changes, but the most ancient sun site in the world is to be found at Nabta in the southern Saharan Desert of Nubia in Egypt. Deserts, which stretch out seemingly endlessly, are ideal locations for observing the sun. Maybe that is why so much early astronomy was developed in flat, arid places endowed with abundant sunshine and clear skies at night. Nabta was

It is not possible to judge precisely when the sun is at its highest simply by looking at it, but with one of the very simplest tools available to early man it is possible to do this and thus find south with accuracy. This tool is often used by sundial makers today, for they must know south precisely to set up their dials. All that is needed is a straight stick positioned vertically on level ground and a piece of string. Throughout a sunny day the tip of the stick's shadow is marked on the earth at a series of intervals. The points so marked will form a curve, which is the path of the sun's shadow. Using the string a circle centred on the stick is then inscribed on the ground to intersect the path of the sun at two points. Using the string again the distance between these two points is measured and halved. The halfway point between the two intersections is precisely north of the stick. The line so formed also points towards the Pole Star, something that must have puzzled early man. No one knows how many thousands of years ago man first discovered this simple technique for finding the cardinal points.

a site where lakes formed in the summer monsoon and was visited over millennia by nomadic herdsmen. In *c.*5000 BC the astronomically minded among these herdsmen constructed a circle of megaliths, like Stonehenge in England, but vastly bigger, occupying an area of about one square mile, and much older, constructed a thousand years earlier than any comparable site in the world. The site has been studied by a team from the University of Colorado at Boulder. Some of the megaliths are nine feet high. An east–west alignment was present between one megalithic structure and two others about a mile away. There are two other geometric lines involving about a dozen additional stone monuments that lead both north-east and south-east from the same megaliths. We do not yet understand the significance of these lines, but the megaliths were probably used to mark the passing of the seasons and thus determine times for planting crops. In Japan, the Kanayama megaliths, dating from 4000 BC, have only recently been carefully studied. Carvings on the giant rocks have been revealed which bear solar alignments.

Later still, other cultures from the Middle East to Central America developed their observations of the sun. In Mexico there are many Mayan sites built for sun measurement. Mayan astronomers measured the seasons to develop a sophisticated calendar and it has been claimed that, before their civilization started to decline in AD 900, they knew the length of the year with accuracy close to the modern value. The Mayans enjoyed a supreme geographical advantage, for they lived in the tropics. The zenith, a word from Arabic which may belie a more ancient Babylonian origin, is that point in the sky immediately above the observer. It is only within the tropics that there will be two moments each year when the sun will be overhead on a vertical line between the observer and the heavens.

It is possible that they used a straight pole set vertically in the ground with a plumb – a string and a weight – which would form a precise vertical line. For only a few minutes, when the sun reached the zenith, there would have been no shadow. Simple instruments can be accurate, and the bigger the better. At Xochilco in Mexico, a pit has been found which has the appearance of a well. Indeed, some have conjectured that it is only a drain, but others have claimed that it is an early zenith tube, and from the bottom of this pit an observer could have observed with accuracy the sun or a star as it passed immediately overhead. The Mayan calendar makers noted the passing of the sun overhead over many years, probably for several hundreds of years, and thereby acquired an enormously clear understanding of all its movements and of the seasons, and built pyramids and temples to demonstrate their respectful and fearful obeisance and admiration.

The Great Pyramid at Giza is the only one of the Seven Wonders of the Ancient World to survive largely intact into our age. Its base is almost perfectly aligned with the four cardinal points. The largest error measures only 5.5 minutes of arc. Each side is 750 feet long and the deviation along the whole of its length would have been about a foot. Such accuracy could have been achieved only by means of the sun.

For western Europeans, the story of measuring the sun starts with Stonehenge in England and New Grange in Ireland. No one knows in full why these immense and wonderful monuments were constructed, but it is reasonable to assert that they were in part inspired by an interest in both science and spiritual life. Britain and Ireland are extremely rich in such prehistoric megalithic structures. It seems that these islands may have contained up to a thousand sites with rings of stone. There were also similar ringed structures made of timber posts, perhaps several hundred of these as well, but most of them have long since been destroyed. Stonehenge was constructed over many centuries starting from 3000 BC. There is clear evidence that it has one alignment towards sunrise at the midsummer solstice and another towards sunset at the midwinter solstice. Neither an observatory nor a sundial, it dates from the time when European

ABOVE A view of Stonehenge at the time of the winter solstice.

LEFT The midwinter sun at New Grange soon after dawn. The roof-box is the opening above the entrance. The door has been left open to illuminate the interior for the photograph.

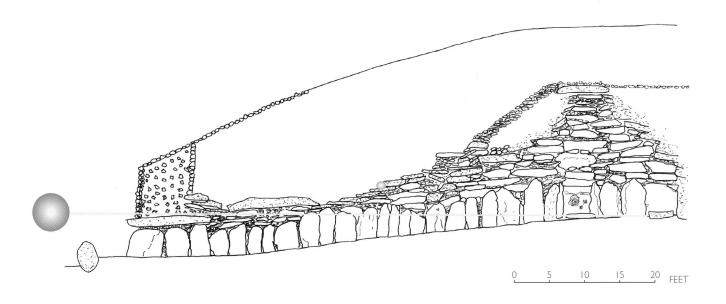

ABOVE A cross-section of the tunnel at New Grange. At dawn on midwinter's day the sun shines down the tunnel through the roof-box .

man was beginning to count the days between each successive solstice, and when astronomy and religion were two branches of the same subject. If you measure the sun in order to worship it, the roles of priest and astronomer must be the same. The setting and rising of the sun at the time of the winter solstice would have marked the death of the last year and the birth of a new one.

Some two hundred years older than Stonehenge is New Grange, a Neolithic tomb thirty miles north of Dublin. It is in a man-made mound, circular in shape and about 280 feet in diameter built with heavy stones up to 12 tons in weight, rubble stones and earth. It forms a tunnel 80 feet long at the end of which archaeologists discovered the remains of several bodies. Many of the stones lining the tunnel have been carved with spirals, zigzags and wandering lines, the significance of which has not been explained. Above the entrance there is an opening – the roof-box. Early in the morning of 21 December 1967, the day of the winter solstice, Professor O'Kelly, the archaeologist who first speculated that the site contained solar alignments, and who had earlier removed some stones from the roof-box which he believed to be a later insertion, walked crouching down the chilly tunnel accompanied by colleagues and waited until dawn. He recorded:

> At exactly 8.54 hours GMT the top edge of the ball of the sun appeared above the local horizon and at 8.58 hours, the first pencil of direct sunlight shone through the roof-box and along the passage to reach across the tomb chamber as far as the front edge of the basin stone in the end recess.

The tunnel was aligned precisely in the direction of the midwinter sunrise. New Grange had been known to antiquaries since 1699 but, until Professor O'Kelly's extensive excavation, the roof-box had been buried under an ancient landslide. The illumination lasted for 17 minutes before the narrow shaft of light through the opening in the roof-box was cut off by the height of the rising sun. The Kelly team may have been the first people for several thousand years to witness this recognition of the death of the old year and rebirth of the sun into its new year. Perhaps the tomb was built to celebrate the death and rebirth of its occupants as well. Maybe it was both a house for the dead and an abode of the spirits. Could the roof-box have been constructed to enable a once-yearly visitation by the sun on the Gods who dwelt inside, or was it a soul-hole through which the spirits of the dead could come and go? The archaeologists have their speculations but do not know why 50,000 tons of stone and earth were accurately placed in a mound that predates the Egyptian pyramids.

EGYPT, GREECE AND BABYLON

Early man had good reason to divide the year into seasons and invent some system for recording date. Indeed, for civilization to progress it was essential for him to develop the skills of a sedentary life as he moved on from nomadism, and to plant seeds and harvest crops at appropriate and predictable times. As yet there was no need to know the time of day. When man first felt the need to divide the day into different intervals and measure them with a sundial is not known. The earliest sun-measuring instruments were only used for observing the sun's seasonal movements or for determining the length of the year. As civilization became more developed, the need arose to divide the day into identifiable intervals, and the first sundials were devised. As will be seen later, sundials come in many forms and can be any shape. The earliest known ones for measuring daytime are oblong, from ancient Egypt and date from the reign of Pharaoh Tuthmosis III in *c.*1500 BC. No doubt employers in antiquity were as keen as ever to get value for money, and it is possible that the need to measure the day first arose to enable the time for different labouring tasks to be calculated. It could also be that these early dials belonged to a high-ranking official or courtier who needed a timekeeper to attend specific engagements throughout the day, or maybe they were of symbolic meaning and associated with the Egyptian sun god.

Some experts think these dials divided the day and the night into twelve hours each, which would suggest that the concept of a twenty-four-hour day is of great antiquity. They were not accurate as we understand it, but they had great symbolic meaning and were a representation of the daily strength of the sun god and the cycle of time. They were used to *define* the time, instead of simply telling the time, as we expect of dials and clocks today.

After the Egyptians there is a huge gap. The sundial disappears from our knowledge for more than a thousand years before it is completely reinvented with fresh thinking. A new and remarkable development, the hemispherical sundial is attributed to a man called Berosus. Like the three wise men of the Christmas story, he was a Chaldean astronomer from Babylon in

modern Iraq. It is known that Berosus moved to the island of Cos in 270 BC and he is considered responsible for transmitting to the Greek world much of the extensive Babylonian records and knowledge of astronomy. He is lucky to have the hemispherical sundial named after him for it is not certain that he invented it, but in the sometimes scant records of antiquity mistakes of this kind were often recorded as fact. For example, everyone has at school either struggled with or been fascinated by the famous theorem concerning the sum of the squares of the sides of a right-angled triangle, named after Pythagoras; but he did not discover or prove it first. It was known at least a thousand years before his time.

Berosus, if it was he, had a simple but brilliant idea. The sky above appears as a giant azure bowl or dome – a hemisphere. He placed a hemisphere on the ground like a kitchen mortar and inserted a pointer into it. The tip of this gnomon met the centre of the circle that formed the rim at the top. The dial was called the hemispherium. As the sun moves across the dome of the sky the shadow of the gnomon's tip will, in reverse, accurately indicate in the bowl the position of the sun in the sky. The mathematics may not have been fully understood at the inspired moment of invention but, when they were, the inside was engraved with a series of curves. Each curve formed part of the Great Circles that can be drawn round any sphere. On the earth, lines of longitude are Great Circles that pass through both poles. On the hemispherium the hour lines are parts of Great Circles. Eleven arcs like this divided the day of Berosus into twelve equal parts, or twelve periods of daylight, each equal in length.

Later, the unnecessary part of the hemisphere, across which the shadow of the tip of the gnomon never passes, was cut away and the truncated result was called a hemicyclium. This is often attributed to Aristarchus of Samos, who was a contemporary of Berosus, but some say it was he who invented the hemispherium and Berosus the hemicyclium. Whatever the truth, Aristarchus was a brilliant astronomer, who was not heeded for all his work. The ancients believed that the sun went round the earth, but Aristarchus believed that the earth went round

Two Egyptian dials from the reign of Tuthmosis III (c.1500 BC). These dials measured the height of the sun. They were probably accurate to less than an hour, but were simple to use. The dials consist of a stone bar at the end of which is a stone block. The bar was turned until the vertical block pointed towards the sun, and the point was noted where the shadow fell on the bar in relation to the five rings marked on it, which are unequally spaced to give approximately equal intervals of time as the shadow shortens. By midday the shadow of the block will be near enough vertical. In the afternoon the block was turned so that it faced the sun in the opposite direction, and the shadows lengthened until sunset.

BELOW A modern hemispherium, 2002, by David Harber, stainless steel, diameter 30 inches. He describes it as a chalice on slate, and the inner hemisphere is a perfect interpretation of the invention attributed to Berosus. The area between the two hemispheres is filled with water, which overflows by means of a circulating pump and is lit to create a ring of light at nighttime.

RIGHT AND BELOW Holker dial, Holker Hall, Cumbria, by Mark Lennox-Boyd. This dial was designed in 1992 by the author for a country house in the north of England and sits on a rock on a small hillock surrounded by open countryside. It is turned from a slab of Burlington slate, 5 feet in diameter, and is marked for every 15 minutes from sunrise to sunset for every day of the year. Gold curves are shown for the hours and for each of the twelve zodiacs, whose signs are engraved on the perimeter. This wonderful piece of craftsmanship is an example of the great skills of slate masons at the Burlington quarry. The dial is derived from the hemispherium of Berosus. It is a projection from the hemispherium on to the segment of a shallower hemisphere, forming a bowl. Such a dial is known as a 'scaphe', from the Greek word for a boat.

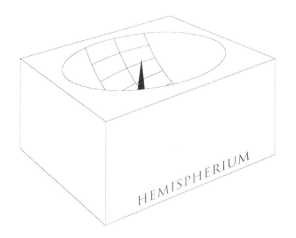

HEMISPHERIUM

SOL OMNIBUS LUCET
The sun shines for all

LEFT This fine hemicyclium is to be found in the ruins of Pompeii and stands elegantly on the top of an Ionic column in the market place, clearly visible to visitors today as it was before the catastrophic eruption of Vesuvius in AD 79.

RIGHT A modern recreation of the sundial that might once have stood in the Roman forum in Leicester, known as Ratae to the Romans. Designed by Allan Mills and sculpted from honey-coloured Clipsham stone, it is 32 inches wide, 29 inches high and 25 inches deep, and weighs nearly half a ton. The bronze gnomon is 12 inches long. This dial may be seen at the Jewry Wall Museum in the city, where it forms part of the Leicester Time Trail.

FAR RIGHT Often additional sculpture was incorporated into classical sundials, in this case two flanking lions. This fine example of a hemicyclium dating from the first or second century AD is in the British Museum.

BELOW RIGHT Hemicyclium, from the temple of Artemis, Ephesus, Turkey. On the base there is inscribed a dedication to the Emperor Severus Alexander (ruled AD 222–35) and Julia Augusta, his mother. The dial is divided into divisions for twelve unequal hours of daylight, which are indicated by the letters of the Greek alphabet.

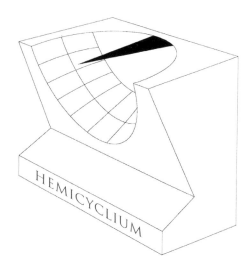

HEMICYCLIUM

the sun, thus anticipating Nicholas Copernicus by nearly 1,800 years. Cleanthes, a Stoic philosopher, even declared that he should be indicted for impiety. Academics often quarrel, and on occasions clever men make profoundly wrong judgments.

The hemispherical dial was further developed by Greek mathematicians. Curves were engraved to indicate date, usually two curves for the summer and winter solstices and one for the two equinoxes. Thus the topmost curve was for the winter solstice when the days are shortest. The middle curve was for the equinoxes when days and nights are of equal length, and the bottom curve was for the summer solstice. No doubt the carving of hemispherical surfaces required skill and time and thus was relatively expensive. To make it cheaper and lighter, the hemispherium was developed into the conical dial. This invention is attributed to Dionysodorus (*c*.250–*c*.190 BC) of Caunus, now a popular tourist resort in southern Turkey. He is known to have studied the geometry of the cone, and so it seems probable that he would have been capable of solving the problem of designing such a dial. It can be understood if one imagines the curves formed by a projection from the tip of the gnomon through a hemicyclium on to a cone that touches it. The central line of the cone is parallel with the earth's axis and points north.

These Greek inventions have occasionally inspired modern architects to cast shadows on to buildings of considerable size. In 1987, Disney World in Florida – 'the sunshine state' – sought a new building in which to consolidate many leased offices. Although it was an office building, it was to be open to the public. The Chairman of Disney wanted it to have entertainment value and a giant cone, standing on end and truncated, was conceived as a sundial by the Japanese architect Arata Isozaki.

BELOW LEFT Time cone sundial, 1997, by Ono Yukio, height 6 feet 10 inches. Ono Yukio is Professor of Art and Design at Tokyo Zokei University and has developed the conical dial into a modern idiom. This sculpture is built of two cones: that of the upright and the part cone of the dial that is sliced by the stainless-steel gnomon so that the illusion of a continuous surface is made by its reflection.

RIGHT Conical dial at Tavagnasco, northern Italy, 2003, diameter *c*.9 feet, height *c*.8 feet. Riccardo Anselmi decided to make a sundial on the roof of a friend's building, which in earlier times had been a store for cheese. It is not a conical dial following the ancient tradition for the dial is depicted on the outside rather than the inside of the cone, and is placed with its axis vertical with the ground, rather than parallel with that of the earth.

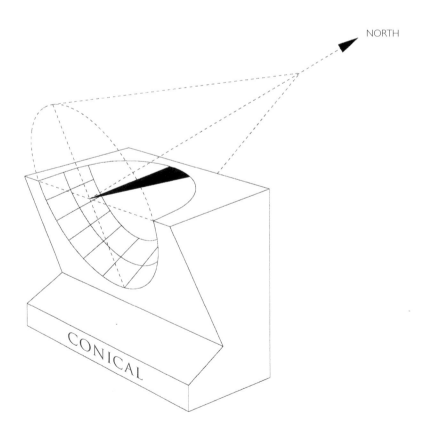

NORTH

CONICAL

The hemispherium of Berosus was mathematically sound; it could be calculated and was developed into other shapes and forms. It was a sophisticated invention that worked by recording the sun's movement across the sky, rather than crudely measuring the sun's height above the horizon as with the Egyptian devices of much earlier date. Its invention was of profound importance for all the sundials that later followed. The earliest known sundial of this type is from the third century BC and the latest from the fourth century AD, and so they had a life of nearly seven hundred years.

If you invent a sundial you have also to decide the basis on which you divide and measure your day. To us it seems self-evident that a day should contain twenty-four intervals of equal duration, hours of sixty minutes and minutes of sixty seconds. Yet these divisions are the inventions of man. The Babylonians used the number sixty as the base for their arithmetic, just as we use ten, and it seems that they were responsible for the division of the hour and the minute into sixty units. The Egyptians divided their day into twenty-four hours and the Greeks put the systems together to give us the system in universal use today. One might add that the Romans gave us the words for minute and second. So our units of daytime measurement are the product of four early civilizations. The word hour comes from the Greek and Latin *hora*, which in turn comes from the ancient Egyptian *har* or *hor* meaning 'day' or 'sun's path'. Horus was the Egyptian god of dawn. Should one hour be of the same length every day? To us, living by the clock, it would be absurd if all hours were not the same length. But before the clock people lived by the sun and in antiquity it seemed more sensible to divide each period of daylight into twelve equal hours. The hours were counted from the moment of sunrise, which was at the beginning of the first hour. Noon was at the sixth hour; sunset was at the end of the twelfth hour. The result was that summer hours were longer than winter hours. Only at the equinox, when daylight is the same

Disney World offices, Florida, architect Arata Isozaki. The building was completed in 1992, conceived as an example of 'entertainment architecture'. Inside there are marble plaques with quotations about time from a wide range of individuals, including Albert Einstein and Donald Duck. The drum, which rises out of the centre of the building, is a truncated hollow cone and there are two gigantic sundials, one visible on the outside and one on the inside. The inner gnomon sometimes casts its shadow for a distance of 130 feet. It is claimed to be the largest sundial in the world.

length as the darkness of the night, were nighttime hours the same length as daytime hours.

In the Mediterranean, where these ideas were developed, the difference between the lengths of summer and winter days is less than it is in northern Europe. So the inequality in the length of the summer hour compared with the winter hour was of far less importance to them. Additionally, the pace of their lives was slower. Dreaded deadlines were fewer, almost non-existent. The intervals of time used in the classical world are called unequal hours. They were in use throughout Europe until an accurate clock was invented, making them obsolete. No one has yet seen fit to rely on a clock that goes at different speeds in summer and winter, night and day.

Greek mathematicians went on to project the curves of the hemispherium on to flat surfaces, both horizontal and vertical. It must have been a delight when they discovered that the time curves were transformed into straight lines. Perhaps the most beautiful of these dials are to be found on the Tower of the Winds in Athens. Fine sketches of this building were published in 1762 by James Stuart and Nicholas Revett in their book, *Antiquities of Athens*. Stuart was an early enthusiast of Greek architecture, which had been ignored for centuries, and helped establish the Greek Revival that spread all over Europe and beyond to Russia and America. As a result of this love he acquired the nickname 'Athenian' Stuart. He was the son of a Scottish sailor and from a poor family. It was not until he was twenty-nine that he managed to get to Rome, a journey which he accomplished on foot. There, he became a respected judge of pictures and found employment as a guide for English visitors. In Rome he met and became friends with the artist Nicholas Revett. Stuart also made the precise drawing of the obelisk of Psammetichus II (see page 37). In 1751 he and Revett travelled together to Athens to record the antiquities. They had considerable difficulty drawing the architecture, for their lives were constantly being endangered by plague and political upheavals and their meticulous methods of measuring and recording the buildings caused them to be regarded as spies by both Greeks and Turks. However, they persevered and one inhabitant even pulled down a house to enable them to get a better view of the Tower of the Winds, though it seems it was rebuilt afterwards. On their return to Britain, they were both elected the first artist members of the Society of Dilettanti, a dining club of English noblemen, gentlemen and amateurs, who met to discuss their love of art during bouts of sometimes drunken hospitality.

The Tower of the Winds was constructed in the market place of Athens by a Macedonian astronomer, Andronicus of Cyrrhus, around 50 BC. It is an octagonal tower of great beauty, about 45 feet high, and remains a famous archaeological site in Athens today. Each face of the building bears a sculpture in relief of a wind god. Underneath the reliefs are sundials, each engraved differently according to its aspect. On the apex of the roof there was originally a bronze weather vane in the shape of a Triton, whose wand pointed to the name of the wind that was blowing. Inside there are the remains of stone channels. It is thought that these were for conducting water to a water clock in the form of a wheel which turned once each day and was decorated with figures of the constellations. Thus it showed which constellations were rising and setting at any moment – even in daytime.

At a slightly earlier date, Greek astronomers in Egypt were at work measuring the heavens and the stars and observing with accuracy the annual movements of the sun. At that time no one knew for certain whether the year was always the same length. The only way to prove that this was so was to measure it. One method of measuring astronomical phenomena accurately with simple instruments is to conduct the observation many times over many years, sometimes over centuries, and average the results to improve the conclusion. Using this method, Greek astronomers measured the length of the year with considerable accuracy.

More than a hundred years before the birth of Christ they calculated the length of the year to within $6\frac{1}{2}$ minutes with a very simple sun-measuring instrument – a metal ring less than a man's height in diameter. That they did so may seem astonishing; their method was clever, but simple and not so difficult to understand. They measured the interval elapsing between one equinox and the next but one a year later. This interval would, of course, be one year in length, and a method was devised for measuring the moment of equinox accurately. The artist's impression shows the sort of instrument they used. It consists of two rings, one inside the other, with both fixed to a frame which is aligned from north to south. The ring that was used for the measurement is the inner ring. The outer ring and the frame are there to fix the inner ring precisely in its correct position, which is parallel with the earth's equator.

To understand what this means, imagine the earth sliced like an apple through the equator. Now think of a ring, somewhere on earth, parallel with this slice. We must imagine ourselves in Egypt, where all this took place. We are far to the north of the equator, and the ring will therefore be at an angle with the ground, and the line of its axis, that is through its centre and perpendicular to it, will point north, in fact to that point near the Pole Star round which the stars appear to rotate at night. The ancient Greeks measured this polar position with great care, and were therefore able to align their instrument accurately. The equinox is not a day, but an instant, and when it occurs the sun is exactly over the equator. An equinox can of course occur when the observer is in darkness, but when it occurs in daylight it can be observed with this simple ring.

Observation (1) by Aristarchus.

Observation (2) by Aristarchus $365\frac{1}{4}$ days later than (1).

Observation (3) by Hipparchus $52,960\frac{3}{4}$ days later than (2).

Note: $(365\frac{1}{4} \times 145) - \frac{1}{2}$ day $= 52,960\frac{3}{4}$ and $\frac{1}{2}$ day $\div 145 = 5$ minutes.

The Greeks did not use our hours and minutes. The figures given are our equivalent of their units.

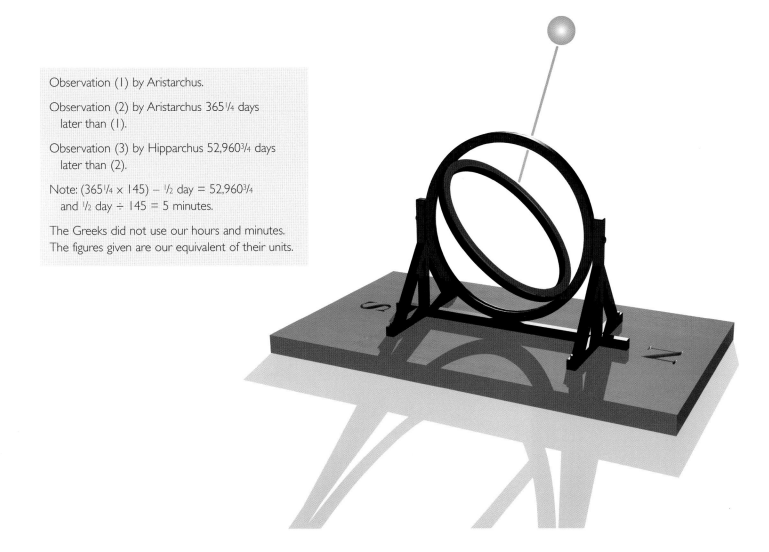

Shortly before and shortly after the equinox an imaginary line drawn from the centre of the earth to the centre of the sun would be above or below the equator. At the moment of equinox this line would intersect the equator, and likewise project along a diameter of this ring.

It follows that at the instant of equinox the shadow of the rim of the ring would fall in line with the opposite inner rim, so that the whole of the inside of the ring would be in shadow. Note, in the illustration, that both the inside and the nearside of the ring are in shadow, and although it cannot be seen the *whole* of the inside is in shadow. This can only happen at the instant of equinox. It was possible to estimate this moment to within much less than an hour. We know of this technique because it is described in the *Almagest*, the great work of the Greek-Egyptian astronomer Ptolemy. This book, known by its Arabic name, was translated from Greek into Arabic in the ninth century and only reached Europe 350 years later.

The story essentially involves two other people, Aristarchus of Samos (whose name appeared earlier) and Hipparchus of Nicaea, who worked for part of his life in Alexandria. In about 275 BC Aristarchus had made two observations of the equinox 365 days apart. He recorded that the second observation was made about six hours later than the first. He probably judged these two times by means of a sundial. He therefore concluded that the year was 365¼ days long. This assertion may well have been a confirmation of observations made by others over time. His calculation was reasonably good but the year then was in fact 11½ minutes shorter than this. A hundred and forty-five years after this second observation, Hipparchus made the same observation himself. He knew the records and dates of Aristarchus's observations and judged that the interval between the second observation and his, the third one, was half a day shorter than it would have been had the length of the year been exactly 365¼ days. So with typical Greek logic he reasoned that the year was 0.5 ÷ 145 days shorter than 365¼ days, or about 5 minutes shorter, which is 6½ minutes longer than it actually was. This story demonstrates how ancient people were able to make accurate measurements with simple instruments, provided they made their observations over long periods of time. Hipparchus had good luck, not least because there were other factors of which he did not know. He was also fortunate that the sun was shining when he needed it but he was not the last person in science to combine good fortune with the imagination and reasoning of genius.

In the *Almagest*, Ptolemy also mentions the armillary sphere, which is of great relevance in the history of sundials, for it gave birth to the armillary dial, and later the astronomical ring dial. It has been used since antiquity to demonstrate the workings of the solar system. The modern example illustrated here is a mobile sculpture showing how the earth orbits the sun. In earlier times armillary spheres showed the earth at the centre. In the model (opposite, below) the earth is on the end of a bar, which swivels on a pin fixed to the sun at the centre. As the bar is turned a pointer at the opposite end indicates the date and the sign of the zodiac – the twelve divisions of the Babylonian calendar year indicated by the constellations that lie behind the sun at any date. The earth rests on a bracket and is fixed to a rod with a weighted end, a bob that keeps its axis vertical. Thus the earth tilts as it orbits the sun. In the illustration (opposite, above) the earth is at the point of summer solstice when the midday sun is at its highest. At the other extreme is the winter solstice, with the two equinoxes in between. The horizontal rings represent the equator, the tropics of Cancer (midsummer) and Capricorn (midwinter) and the Arctic and Antarctic circles. We are familiar with these on the globe of the earth, but to astronomers and in the model they are projected on to the celestial sphere and are represented by these rings which one must imagine as projected infinitely into space. The tropics of Cancer and Capricorn are so called because the sun projects on to them at those times of the year when the constellations of

RIGHT A diagram showing another way of understanding how the earth orbits the sun. At the summer solstice the earth is tilted towards the sun, and at the winter solstice in the opposite direction. At the moment of equinox an imaginary straight line from the centre of the sun to the centre of the earth will intersect precisely with the earth's equator.

BELOW RIGHT An armillary sphere, 2003, by Mark Lennox-Boyd, painted wood, paper and steel, diameter 32 inches. Models of the heavens such as this have been in use for nearly two thousand years to demonstrate man's understanding of the geometry of the solar system. The earth is here positioned at the time of the summer solstice. The small horizontal rings at the top and bottom represent the Arctic and Antarctic circles, projected out from the earth on to the celestial sphere, a great globe of crystal that the ancient Greeks imagined enveloped the earth. They were wrong. Of course there is no crystal sphere and the sun, not the earth, is at the centre of the solar system, but we still use the term 'celestial sphere' to describe the dome above us that we see at night. The model is correct, with the sun at its centre. The central band represents the equator, the band above the tropic of Cancer; below the equator is the tropic of Capricorn, all projected on to the celestial sphere. The diagonal band between the tropics is the ecliptic, in which the earth orbits the sun. The ecliptic is divided into twelve equally spaced sectors – the zodiac – the dates of which are depicted against their appropriate signs.

In Cairo, in the early eleventh century, an armillary sphere was constructed fitted with nine rings, each weighing about a ton, and each large enough to ride through on horseback.

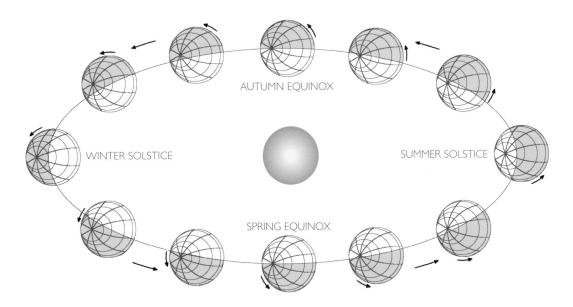

AUTUMN EQUINOX

WINTER SOLSTICE

SUMMER SOLSTICE

SPRING EQUINOX

LEFT A modern armillary dial, 2002, by David Harber, stainless steel, diameter 39 inches. The time is read from the shadow of the diagonal rod cast on the slanting semi-circular band. The rod is therefore the gnomon and the band in the position of the ecliptic band on an armillary sphere has the times of day engraved upon it. The other rings and bands (for the horizon, equator, Arctic, Antarctic, Cancer and Capricorn circles) are merely decorative and are included for historical reasons. They follow the rules for an armillary sphere. The reflective quality of the plinth enables spectacular optical illusions to be created.

ABOVE The images of the zodiac are of great antiquity. Sagittarius, half horse, half man, with his bow and arrow, is shown with a little Scorpio. A man is standing on the back of Capricorn, a goat with the tail of a fish. Another man is standing behind a Leo. Another figure, representing Aquarius, is pouring water from two flasks over one of the two fishes of Pisces. Capricorn and Leo are Babylonian and date from c.1200 BC. The other two images are Graeco-Egyptian from AD c.25.

Cancer and Capricorn lie in the celestial sphere behind the sun. The bar rotates on the sun so that the earth's orbit follows the ecliptic, an imaginary band across the heavens which is so called because it is that part of the sky in which the moon eclipses the planets, the stars and the sun. The twelve signs of the zodiac are part of contemporary myth and our popular culture. They are often depicted on sundials for historical and decorative reasons. Few in the west believe for certain in the predictions accorded to our stars, but millions want them to be true and few of us can resist an opportunity to read them.

It is worth reflecting that a majority of people in the East still believe in astrology, and certainly for a great deal of the last 2,500 years most people in the West did so as well. Even brilliant thinkers adopt unusual and unexpected beliefs. In the seventeenth century the greatest man of science and reasoned thought in the world was Sir Isaac Newton. It was said of him by a contemporary that his contribution to mathematics was greater than the total of all mathematicians who had preceded him. But much of his work was conducted without publication, certainly his work on alchemy. He was a prodigious writer on every subject that he studied, and wrote in secret more on alchemy than on mathematics. Many of his contemporaries were astrologers. The mystery of the stars was very disturbing before mankind knew much about them. While most stars moved along fixed paths, the planets appeared to be stars with erratic motions, seeming to wander backwards as well as forwards. The Babylonians called them the stray sheep. Even when a measure of understanding was acquired it was difficult not to imagine that the presence and unpredictable movement of these heavenly bodies somehow had meanings which affected the destiny of people on earth.

According to the principles of Renaissance astrology, the seven known planets in the order of their decreasing distance from the earth were:

Saturn, Jupiter, Mars, Sun, Venus, Mercury, Moon

According to this theory, the sun and the moon were both treated as planets. The planets were understood to possess individual characteristics and to rule different hours of the day. The planet that ruled the first hour of a day had a special astrological significance and power over the day itself. Mindful of the decreasing distances, they assigned the first hour of the first day of the week to the sun, the second hour to Venus, the next planet, the third to Mercury and so on, repeating the above cycle every seven hours for twenty-four hours. If you count the hours against the cycle you will see the twenty-fifth hour, or rather the first hour of the second day, was ruled by the moon, and the first hour of the third day by Mars. The planets thus gave their names to the days of the week in the Latin languages. Anglo-Saxons substituted four northern Gods (Tiw, Woden, Thor, Freya) for Tuesday to Friday, for it was believed they corresponded to the Roman ones.

We must all answer whether it is wrong to believe that mankind's condition and fate are determined by the same celestial motions. No one disputes that these fix the weather, the tides, the seasons and so the behaviour of some animals and the success of a harvest. The menstrual cycle on average is the same as the interval between one new moon and the next. So maybe the fates and characters of men are affected by the heavenly bodies? Many believe this. Some of the stars were seen to form groups, twelve of which appeared to encircle the earth in a vast ring and were seen against a fuzzy band, which came to be called the Milky Way. The planets also appeared to occupy this part of the heavens and at night the moon passed through it and was occasionally seen to eclipse a planet by passing in front. So it came to be believed that the twelve groups of stars and the planets moved around the earth in a giant circular band – the ecliptic.

The Babylonians inhabited a part of Iraq where the nights were sharply clear provided the desert winds did not whip up a dust storm. In about 500 BC they were the first to measure systematically those groups of stars which the Greeks later called the zodiac. The word means a small animal, for it had long been observed that some of the groups appeared to form shapes similar to such images as scorpions and crabs. The Babylonians used the zodiac of 12 × 30° (360°, the circumference of a circle) as a reference system of solar and planetary motion. It is this system that is sometimes seen as decoration on sundials, for it is closely related to date. The measurement started at the spring equinox, known as the first point of Aries (0°), normally 20 March. On this day the sun entered Aries, that is to say it lay on a straight line drawn from the constellation to the earth. Of course one could not see this because of the brightness of daylight, but one could infer the fact by seeing the opposite constellation in the circle pass south at midnight. And so the sun moved through three constellations until the summer solstice at the start of Cancer (90°), 21 June; and then to the autumnal equinox at the start of Libra (180°), 23 September; to the winter solstice at the start of Capricorn (270°), 22 December; and finally back to Aries. Some of these still popularly known symbols date back more than 3,000 years.

The only problem about using this decoration on sundials is that the twelve constellations have been shifting their positions since the Babylonians first recorded them. So the zodiac must be seen as decoration for sentimental, historical and of course aesthetic reasons for their images can be very beautiful. In the last 2,500 years each constellation has moved by about 30°. The reason for this is that, as it spins on its axis, the earth wobbles like a toy top that is slowing down. As it revolves each day, its tilted axis is itself turning so that the North Pole is rotating in a separate circle, though much more slowly than the earth's rotation, in fact making one revolution in 26,000 years. This causes an apparent change in the positions of the stars. The motion was known in ancient Greece but it required Newton to demonstrate that the wobble was caused by the moon's irregular gravitational tug on the earth. It is referred to by astronomers in a delightful and grandiloquent phrase of some antiquity, the Precession of the Equinoxes.

Sundials can be described as a footnote in man's study of the cosmos. They have been the basic instruments for measuring the path of the sun. Astronomy, which included astrology, had its birth in those parts of the Middle East between the rivers Euphrates and Tigris – the fertile crescent – from where Western man also first learnt of sedentary living and planting seeds; it was then enhanced in Egypt and greatly by the civilization of ancient Greece, where the accumulated knowledge matured into a body of understanding, much of which would remain unquestioned for more than fifteen hundred years.

RIGHT The twelve images of the zodiac, as represented in the delightful marble inlays on the floor of the great church of Santa Maria degli Angeli in Rome. The table shows the zodiacs, their signs, the corresponding solar longitude, and declination of the sun. Solar longitude is measured from the start of Aries, known by astronomers as the first point of Aries, the date of the spring equinox. It is the distance in degrees that the sun can be imagined to sweep during a year across the sky against the background of the (invisible in daylight) constellations. Declination is the tilt of the earth's axis in relation to its orbit round the sun. At the summer solstice the earth's axis tilts 23.45° towards the sun, and at the winter solstice by a similar amount in the opposite direction. At the equinoxes the declination is zero.

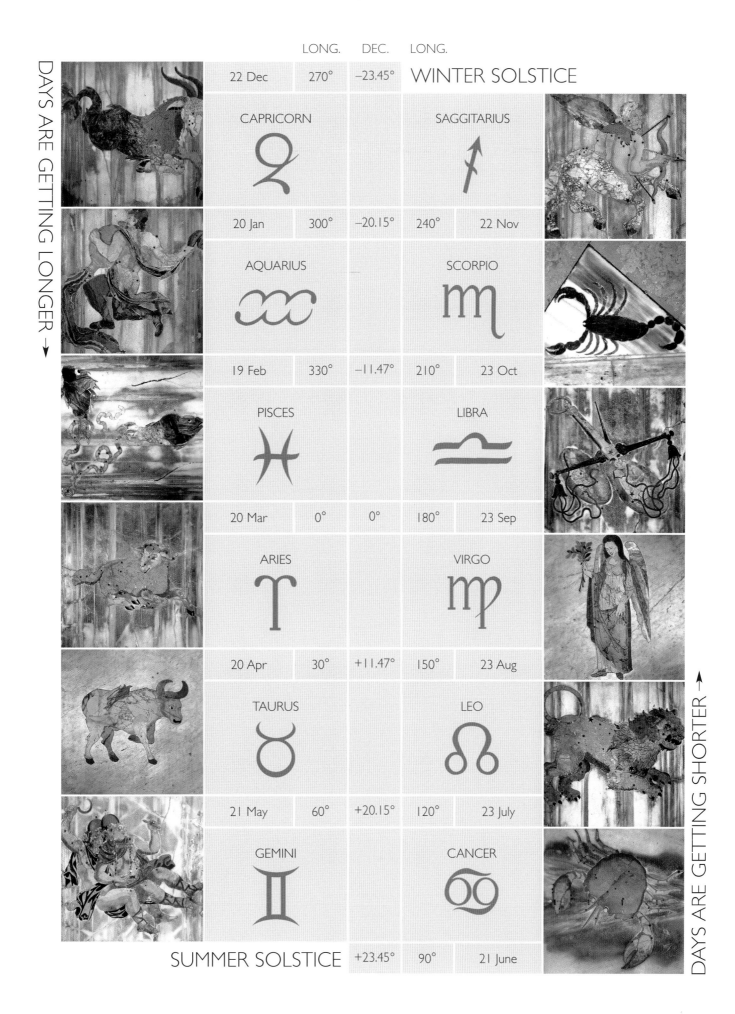

DAYS ARE GETTING LONGER →

DAYS ARE GETTING SHORTER →

	LONG.	DEC.	LONG.	
22 Dec	270°	−23.45°		WINTER SOLSTICE
CAPRICORN				SAGGITARIUS
20 Jan	300°	−20.15°	240°	22 Nov
AQUARIUS				SCORPIO
19 Feb	330°	−11.47°	210°	23 Oct
PISCES				LIBRA
20 Mar	0°	0°	180°	23 Sep
ARIES				VIRGO
20 Apr	30°	+11.47°	150°	23 Aug
TAURUS				LEO
21 May	60°	+20.15°	120°	23 July
GEMINI				CANCER
SUMMER SOLSTICE		+23.45°	90°	21 June

ROME

CUI DOMUS
HUIC HORA

*Each household
should have the time*

Greek science was later absorbed into the Roman empire, and the plane dial was developed even further. Aquileia is in north-east Italy, not far from the border with Austria and former Yugoslavia. It was a frontier town. The Aquileia dial was carved in the second century AD on a stone table surrounded by three benches. It is made up of eleven hour lines, which are drawn from the curve of the summer solstice and which are extended, after being crossed by the equinoctial line, to that of the winter solstice. The shadow of a vertical pointer would follow the upper curve on midsummer's day and the lower curve at midwinter. At the equinox the shadow would follow the broken line. The whole design is enclosed in a circle and marked with the names of eight winds, and the maker's name is added: Marcus Antistius Euporus. It appears that the area of discovery was the site of a circus, which indicates the importance of the city. One can imagine a group of friends sitting round this dial, lingering over their midday meal after some jugs of wine, made vividly aware by the dial that the afternoon was beginning to slip away – that for all their conviviality, time marches on. This image has of course always fascinated mankind. They perhaps reflected, as did the seventeenth-century English poet Andrew Marvell, who was himself fascinated by Roman civilization, in his poem 'To His Coy Mistress' that 'we cannot make our sun stand still'. For apart from their practical purpose in time-telling, sundials offer us a very graphic representation of the passing of time. Aquileia was a large and strong fortress until Attila the Hun swept down upon Italy and left it a heap of ruins in AD 452. It had been the chief city of Venetia until barbarian invasions caused the citizens to seek refuge in the marshes of the lagoon, and thereby indirectly gave birth to the glories of Venice.

Vitruvius, the first architectural historian, who lived in the age of Augustus, described the dials that were then known. These included hemispherical, conical and plane dials. He mentioned the Tower of the Winds. In his monumental work *De architectura* he presents an introduction in which he laments that authors are not accorded the same honours and salaries as professional athletes, a cry that has echoed through the ages, and would certainly be valid today. Vitruvius believed that

LEFT Detail of the dial of M. Antistius Euporus.

BELOW The village of Solimbergo di Sequals is close to Aquileia and a circular piazza there has been enhanced with a sundial inspired by the famous Roman seat dial. The dial is in the centre of a grass circle and the plaques on the five uprights have engravings of the zodiacs.

RIGHT Sundial of M. Antistius Euporus, Museo Archaelogico, Aquileia.

the architect should be equipped with knowledge of geometry, history, philosophy, music, medicine and astronomy, as well as building construction. So he included descriptions of sundials.

Many lovely sundials survive in the museums of the former Roman world, but the triumph of them all must be the great creation of Augustus, tantalizingly invisible to the modern world but recently excavated in several places (see pages 36–7). This monument still lies under the teeming streets of Rome, and would have been trampled by the feet of countless Roman soldiers, for it was built in the Campus Martius, the parade ground named after Mars the god of war, and covered an area of more than 6,000 square yards. Time and date were read from the shadow of the tip of a gnomon nearly 100 feet high. This was the obelisk of Psammetichus II, the first of the many Egyptian obelisks to be removed to Rome. Augustus took it as war booty after the defeat of Cleopatra and her lover Mark Anthony at Actium in 31 BC and had it made into the gnomon of his sundial. On the base is a Latin inscription, which translated reads in part: 'Emperor Augustus, son of divine Caesar, dedicated the obelisk to the sun when Egypt had been brought under the sway of the Roman people.' It was a considerable achievement to get the massive stone back to Rome intact and Augustus had the ship that carried it laid up in the harbour near Rome as a tourist attraction. It came from Heliopolis, City of the Sun, in Egypt. The Emperor associated himself with the sun, and the obelisk was a visual symbol erected to demonstrate his power.

In 1748 it was excavated and re-erected with difficulty, for it had been greatly damaged by fire, possibly when Rome was sacked in 1084. By coincidence, James Stuart was then in Rome on his way to Athens. He wrote a treatise on the obelisk in Latin, which he had learnt in Rome, and was commissioned by the Pope's Secretary of State to make accurate and beautiful drawings, which were published in the official report of the archaeology. The obelisk stands today in the Piazza del Parlamento in front of the Italian Parliament building less than 55 yards from its original location, and from its tip it still tells the time of noon against a modern bronze line set into the pavement.

Augustus was born on the date of the autumn equinox. The straight line marking the equinox points eastwards precisely to the *Ara Pacis* – the Altar of Peace. This contemporaneous monument had been constructed to symbolize his achievements and the new era of peace occasioned by his immensely successful reign, during which he had extended the frontiers of the empire and brought stability and prosperity to the people who were profoundly unsettled after the assassination of Julius Caesar. It is claimed the obelisk is also orientated to symbolize the completion of his life. The northern edge points a little to the west and directly towards the gigantic Mausoleum of Augustus (off the plan on page 36). Such was his ambition and confidence in his future success that the Emperor had had this commissioned eleven years earlier, when he was aged only twenty-nine.

The dial itself, described by the Roman historian Pliny, and so known of during the Renaissance, was rediscovered early in the sixteenth century and a contemporary account

RIGHT Sundial Bench 2, 1994, by Ono Yukio, length 9 feet 10 inches. A sundial can be depicted on any surface that receives the sun but, on a curved surface, straight lines are projected into curves. Ono Yukio has made a sundial bench reminiscent of Aquileia but popularized in a modern playground. This designer has concluded that a contemporary sundial need not be an accurate timekeeper, but rather an object that demonstrates the harmony of the mathematical curves.

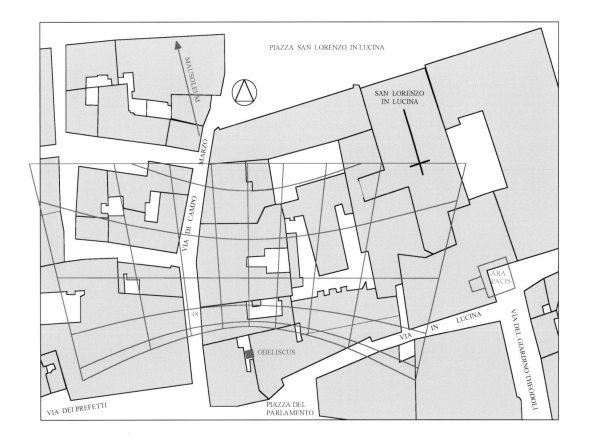

In the map:
PIAZZA SAN LORENZO IN LUCINA

MAUSOLEUM

SAN LORENZO
IN LUCINA

VIA DI CAMPO MARZO

ARA
PACIS

VIA IN LUCINA

VIA DEL GIARDINO THEODOLI

48

OBELISCUS

VIA DEI PREFETTI

PIAZZA DEL
PARLAMENTO

SOL REX REGUM
O sun, king of kings

ABOVE In the sundial of Augustus, the obelisk (*obeliscus* in Latin) told time from the shadow of its tip and was placed so that its northern face pointed towards the mausoleum of Augustus. The entrance to 48 Via di Campo Marzo is marked in red.

describes how its remains were unearthed by a baker digging a latrine. Pope Julius II had no funds to spare, so the remains were reburied and remained so until they were rediscovered for a second time by German archaeologists led by Dr Edmund Buchner, who examined the site. Inevitably they were not able to dig extensively because of the overlying buildings. Shafts were dug in several positions in the surrounding streets and courtyards, and scaffolding erected underground to ensure there was no subsidence near by. In a cellar of 48 Via di Campo Marzo the digging conditions were bad. The cellar was flooded because the water table had risen over the past two thousand years and the ceiling had to be propped up to prevent subsidence, but some thick and magnificent bronze inlays of lines and Greek lettering were discovered at a depth of 24 feet. Roman science came from the Greeks, and many scientific matters were written in Greek.

The illustration above shows the plan of the dial according to Dr Buchner's hypothesis (some archaeologists disagree). Readings were taken from the tip of the obelisk. The straight lines show unequal hours; the central horizontal one is the line followed by the tip on the days of equinox. The curves show the position followed by the tip at different dates, signified by the signs of the zodiac, the lowest curve being for the summer solstice (Cancer and Gemini) when the sun is at its highest.

This great astronomical instrument, one of the most remarkable monuments of imperial Rome, symbolizes the defeat of Egypt, the achievements and the power, the birth and death of the greatest of Roman emperors and the peace and prosperity he inaugurated. It is surely an artistic and scientific statement of imperial ideology unequalled in history.

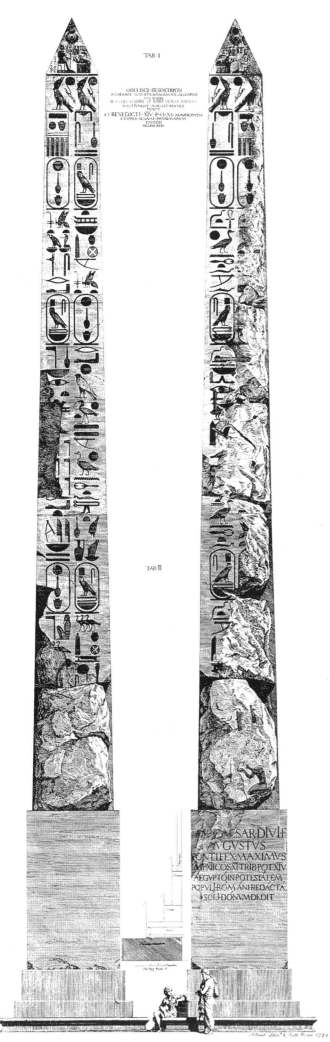

LEFT *The Obelisk of Psammetichus II,* engraving by James 'Athenian' Stuart in *De Obelisco Caesaris Augusti* (A.M. Bandini, Rome, 1750).

BELOW Edmund Buchner had calculated where to dig and he knew the depth at which the engravings would be if his conjecture were correct, because the depth of the base of the obelisk had been recorded earlier. The illustration is a line drawing of what he found. The bronze Greek letters are large, 10 inches high and ¾ inch thick, inlaid into stone. The vertical line is noon, and the short horizontal ones the divisions for each day. To the left, inlaid vertically down, are the Greek letters Π, Α, Ρ, Θ, a truncation of the word *Parthenos,* the Greek word for Virgo. To the right inlaid vertically up are Ο and Σ, the last two letters of *Krios,* the Greek word for Aries. Towards the top below a horizontal line are two words that signify the start of Virgo. The horizontal line above marks the equinox.

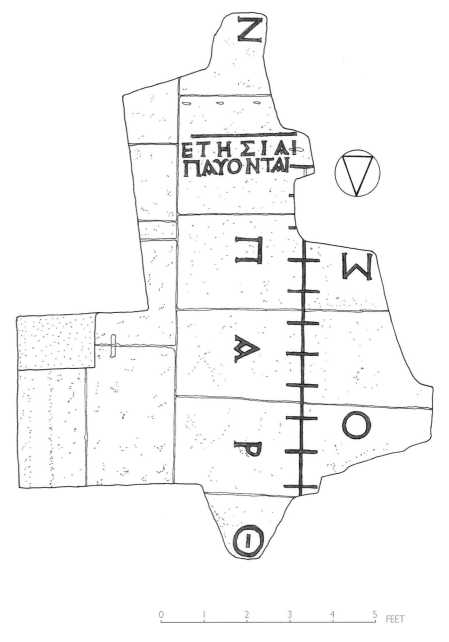

0 1 2 3 4 5 FEET

4 ISLAM AND THE MEDIEVAL CHURCH

When the Arab empire swept into the former Roman world the knowledge their scientists had discovered was seized with enthusiasm and developed with new understanding. Many of our words of science and astronomy have Arabic origins: zero, algebra, sine, almanac, azimuth, nadir, zenith, to name a few. We know that by the middle of the thirteenth century Arab astronomers had introduced the use of hours of equal length day and night, summer and winter, as well as the sloping gnomon. When one looks at the quintessential English garden sundial it is strange to imagine that its design is derived from the invention of medieval Arab astronomers.

Europe was ignorant of these developments; the Dark Ages had descended and Greek and Roman science had not returned to its native continent. Sundials were used in western Europe, but they were often crude and unsophisticated. Time was in the hands of the medieval Church, and daytime was divided into periods relating to the hours of canonical prayer. These canonical hours were derived from the unequal hours of the Roman world and were suitable for religious practices. They had been codified by St Benedict in the sixth century after his foundation of the monastery of Monte Cassino, south of Rome, and defined the following times: 'prime', said at the first hour of daylight, 'terce' at the third, 'sext' at the sixth, 'none' at the ninth and 'vespers' in the last hour of daylight; 'compline' was said after dark and 'matins' before dawn. Later 'none' was moved to midday and gives the English language the word 'noon'. The phrase 'high noon' exists in English because 'none' was originally the ninth hour and at midday the sun was higher in the sky than at noon.

Canonical hours and numbered hours were often used in tandem. The birth of the future Richard II of England is recorded variously to have occurred 'at the time of terce' and 'when it was 11 hours'. In monkish circles the times of prayer were often ill-defined. The dials at Chartres and Bewcastle in Cumbria do not even measure equal intervals of daylight. Canonical hours were also sometimes referred to as seasonal, temporal or planetary hours. In medieval Europe many technical terms were described differently in different places. There was no standardization. But

OPPOSITE Chartres cathedral dates from the twelfth century, though the great sundial has no doubt been recarved several times. It tells a form of unequal hours, known in medieval Europe as canonical hours. In the background can be seen a noon mark. These marks are described in the next chapter.

THIS PAGE In the beautiful landscape of Cumbria, in the churchyard of Bewcastle there stands this monolith, a shaft nearly 17 feet high, which has lost the cross that adorned its head. It is a surprise to any visitor to find such an object in this lonely place. It bears several inscriptions in runes and on the south side the words 'FRUMAN GEAR – KÜNINGES – RICES THAES – ECGFRITHU' (First year of the King of this realm Ecgfrith). Some way further up, about a quarter from the top on the face to the side of the standing figure is a semi-circular sundial between panels of foliage; once it had a horizontal gnomon. The Anglo-Saxon King Ecgfrith reigned from AD 670, and the shaft bears the earliest English sepulchral inscription (in memory of King Alchfrith) and the earliest English sundial. The style of its carving is Byzantine and according to the Venerable Bede Archbishop Theodore's mission took place at this time. He was a Greek and taught the English both the gospels and the 'arts of astronomy and arithmetic'. It could be that this early foreign bearer of the Christian message was connected with this dial.

the situation is yet more confusing. Some communities ran a system of twenty-four *equal* hours from one sunset to the next. These were called Italian hours. They began at zero and ended at twenty-four at sunset the following day. So daylight hours were numbered from about 10 to 24. Babylonian hours, in contrast, ran from one sunrise to the next, and so dials numbered in this way were usually calibrated from 1 to 14. They too were hours of equal length, and markings of these three different kinds of measurement can be found on medieval dials from all over Europe and the Arab world. An early book on sundials with illustrations is *Rudimenta mathematica* (1551) by Sebastian Munster. It contains a fine woodcut in which much of this and an explanation of the zodiac is portrayed.

However, all this was yet to come in Europe. The science was kept alive by Islamic astronomers from North Africa to central Asia.

In some ways, the designing of sundials can be compared to the engineering of great bridges. They are the products of mathematics applied to a functional need, and the results are often of great visual harmony and beauty. The constraint of mathematics often makes good design for the imagination is fired by the limitations of the medium to discover new ideas which still obey the rules, just as trees produce their finest timber when grown at the margins of their environment. Muslims were normally forbidden to represent the human image because of the Prophet's fear they would depict religious images and come to worship the image rather than Allah. Their artistic impulses combined with a love of science drove them to find expression in the design of complex arabesques and architecture. This led to the study of gnomonics and perhaps the greatest example of this knowledge is to be found in the Umayyad mosque of Damascus. This dial was designed in 1372 by a *muwaqqit* whose name was Ibn al-Shatir.

The Arabic word *waqt* means 'time' and *muwaqqit* means 'timekeeper'. This word first appears in the thirteenth century and implies far more than the literal translation suggests. The holder was a distinguished professional astronomer who was associated with a religious institution. One of his duties was to regulate the prayer times, of which there are five, but he also constructed instruments, wrote treatises on astronomy and gave instruction to students. The Abbasid caliphate had ended some years before the Damascus dial was made, but it was during their rule and patronage that Greek science had been absorbed into Arab civilization and developed. This sundial therefore probably represents the culmination of their sophistication in mathematical design. It is also an image of piety in its references to the times of prayer prescribed by Islam.

In Islam, as in no other religion in human history, the observance of various aspects of religious ritual has been in accordance with scientific procedures. The Islamic calendar is strictly lunar and its organization, together with the regulation of the astronomically defined times of prayer, are topics of traditional Islamic science which have a history that goes back nearly fourteen hundred years and are still of concern to Muslims today. For scholars of the sacred law, the month began with the first sighting of the crescent moon. The start of the month varied from place to place, as the moment of first appearance differed in different places. The sighting is easy enough if the weather is fine, you know the direction to look – the western sky – and the approximate time. Witnesses with exceptional eyesight were sent just before sunset to locations that offered a clear view of the western sky. Their sighting determined the beginning of the month. At that instant they shouted 'hallelujah', or rather, the Arabic equivalent of the Hebrew phrase, '*ya hillala*', which means 'behold the crescent moon'. (The Arabic also means 'to rejoice in acknowledging God'.) Like the sun, the moon moves across the sky from east to west, so the thin and faint crescent of a new moon is only visible in the western twilight and for a few moments before it sets. All this explains why a waxing moon has the shining part facing west while the bright part of a waning moon faces east.

TEMPUS VOLAT,
HORA FUGIT
Time flies, the hour escapes

PER SEBASTIANVM MVNSTERVM·

This engraving from Sebastian Munster's book *Rudimenta mathematica* (Basel, 1551) shows almost everything then known to dial makers. The woodcut illustrates a dial facing due south whose gnomon would be a slanting rod emerging from the centre of the sun image. Around this image is a circular calculator, which was used for fixing the date of Easter.

Equal hours are shown by the lines, which bear Roman Gothic numerals and converge on the sun image. The garland also indicates equal hours in Arabic numbers. Italian hours are shown by lines numbered from 14 to 24; likewise

Babylonian hours are shown from 1 to 10. The unequal hours are also shown by lines which bear no number. In this illustration, they are harder to make out.

The date is shown by month in Latin and by zodiac with images and astrological signs. The central line is the equinox, the top curve the winter solstice when the sun crosses the tropic of Capricorn, and the bottom curve the summer solstice when the sun crosses the tropic of Cancer. All dates are read from the shadow of the tip of the gnomon. To the sides of the dial lines are three columns which show:

• The number of hours of daylight and darkness for any date.

• The times of sunrise and sunset.

• The planetary house (astronomical sign for a planet) which rules each zodiac sign in astrology.

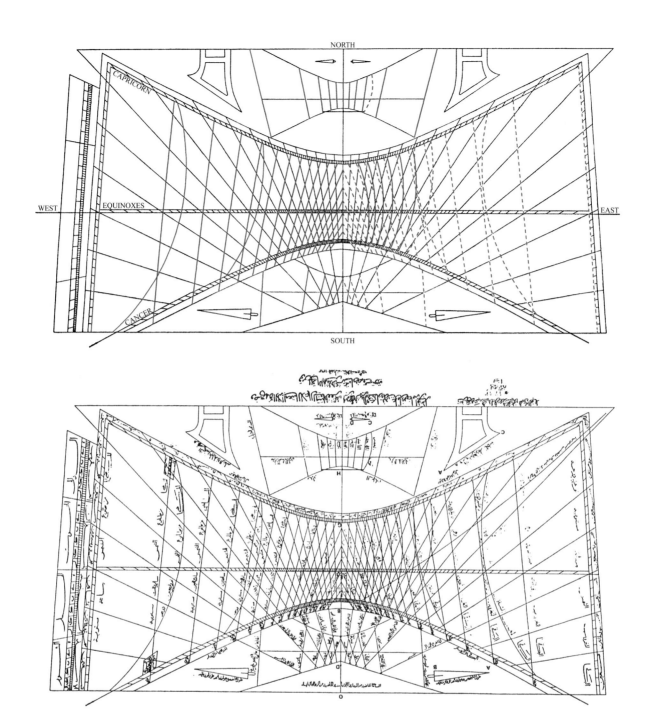

Illustrations of the Umayyad mosque dial at Damascus. The dial is a faithful copy of the original, which was shattered at some time in the nineteenth century, and is now housed in the museum. It is large, measuring 6 feet 7 inches by 3 feet 3 inches, and a masterpiece of complexity. The top illustration has been simplified by the removal of the Arabic text and the rendering of the design in different colours to explain the significance of the detailed engravings. The dial

is provided with a slanting gnomon (probably one of the earliest ever designed) in two parts which nearly join, and it is from the tips of the shadow of the gnomon that many of the readings are made. The curves for the two solstices, the line for the equinoxes and the conventional time lines are shown in black. The main divisions of the time lines are for every twenty minutes, in the equal hour system; the minor divisions divide these by five and

so are for every four minutes of time. But in Damascus in the fourteenth century the concept of sixty minutes in an hour was not in use, and it is more accurate to say that each of these minor divisions represent 1° of hour angle, a term which is explained in the Appendix (see page 128). There are three dials. On the small northern dial, in light blue, time lines are shown following the system of time measurement of unequal hours. The solid red lines

on the central dial show time measured every twenty minutes from dawn (Babylonian hours) and the broken red lines show twenty-minute intervals remaining until sunset (Italian hours). The same lines are shown on the south dial for the hours only. But the originality of the dial is in the times for prayer. In Islam, the pious are required to perform their prayers (*salat*) five times a day, and the times are explained in the captions opposite.

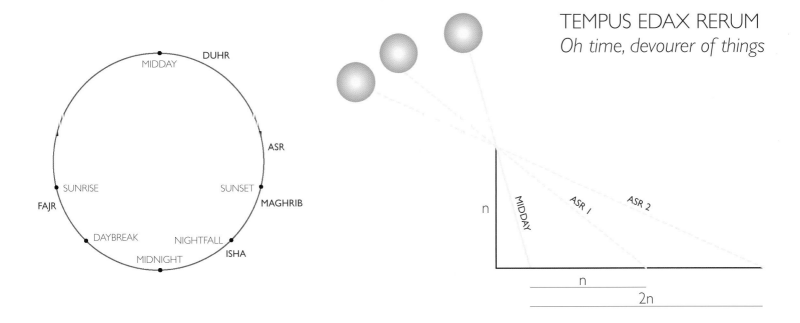

ABOVE AND OPPOSITE In Islam, prayer times are closely related to the movement of the sun. The diagram above right illustrates how the height of the sun determines the start and end of the Asr prayer.

MAGHRIB This prayer commences at sunset.

ISHA This prayer commences at nightfall and the broken red curves show twenty-minute intervals in equal hours, which must elapse before the moment for the prayer.

FAJR The dawn prayer. The solid red curves show the intervals that have elapsed since the commencement of this prayer at daybreak.

DUHR The post-midday prayer. The divisions for this prayer are shown by six solid curves in green.

ASR The afternoon prayer. The commencement of the prayer times is shown by the most easterly curve in green which is broken. A similar curve is shown on the northern dial.

It need hardly be added that Ibn al-Shatir wished to demonstrate that he could combine his mathematical and artistic intellects under the mantle of religious devotion.

The Islamic day is considered to begin at sunset because the months begin when the new moon is seen for the first time shortly before sunset. The first prayer is the *Maghrib* at sunset. The word means 'west' and is also used to denote those Arab countries in the western Mediterranean. The second prayer is the *Isha* or evening prayer at nightfall, 'when the eyelids of the sun have closed'. The third is the *Fajr* or dawn prayer, which begins at daybreak, 'when the eyelids of the sun are opening'. The fourth is the *Duhr* or noon prayer, which begins shortly after midday when the sun has passed south. The fifth is the *Asr* or afternoon prayer, which is the most interesting to sundiallers. The time for this prayer begins when the shadow of any object has increased beyond its midday minimum by an amount equal to the length of the object casting the shadow, and ends when the length is twice as great. The hours normally used by medieval Arabs were unequal. The prayer times corresponded with the times of the seven prayers of early Christianity, but with the dropping of a midmorning prayer and the omission of a prayer at sunrise which was expressly forbidden by the Prophet for fear of encouraging sun worship. Thus there are parallels with the canonical hours of Christianity and indeed the Islamic prayers derive from them. One method of regulating the daytime prayers was to use a sundial. It is reported that the Caliph Umar, known as the Pious, who refused to enter the church of the Holy Sepulchre in Jerusalem for fear that a fanatic would one day make it a mosque, used a Graeco-Roman sundial, probably a hemispherium, marked with the seasonal hours to regulate his prayers. Once, there must have been thousands of Islamic mosque dials. Only a few survive to this day.

While Europe languished in intellectual darkness, the science of Islam flourished even further to the east in central Asia. The name Tamerlane suggests a conqueror with little time for more intellectual pursuits. More accurately known as Timur the Lame, he had built an empire which comprised the areas now occupied by Uzbekistan, eastern Turkey, Syria, Iran, Iraq and India as far as Delhi. He died in 1405 leading his armies into China. His grandson Ulugh Beg, however, is remembered as a hero in Uzbekistan for his remarkable contributions to astronomy. His sextant sundial was the largest ever built. Shortly after Tamerlane's death Ulugh became his father's deputy and at the age of sixteen was made ruler of Samarkand where he built his remarkable observatory. To his ultimate cost, he was throughout his life far more interested in making the city a cultural centre than he was in politics or military conquest. Though he did not neglect the

ABOVE A commemorative postage stamp issued by the government of Uzbekistan in 1994 to celebrate the 600th anniversary of the birth of Ulugh Beg.

LEFT The remains of the great sextant of Ulugh Beg in Samarkand.

 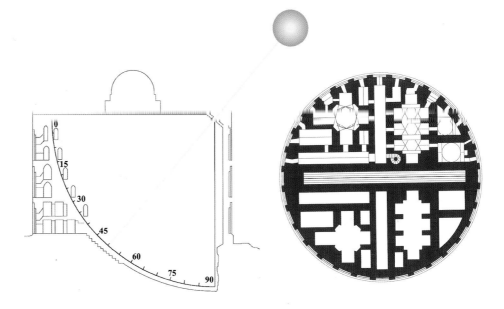

arts he was primarily a mathematician and astronomer and he attracted to his court more than sixty scholars, whose work produced some astonishing results. You may remember the constant for π, the ratio between the diameter and circumference of a circle, as about 3.14. Ulugh Beg's Astronomer Royal calculated this value to sixteen decimal places.

His observatory had many instruments, of which the sextant was the most accurate. It measured the height of the sun in angles as small as the diameter of an American penny at 1,600 feet. Data from this instrument allowed him to calculate the length of the year as 365 days, 5 hours, 49 minutes and 15 seconds, about one minute out and a value not improved upon for nearly three hundred years.

Unfortunately, when it came to political responsibilities, he neglected matters or was incompetent. He died violently, murdered at the instigation of his son. When his tomb was discovered in 1941 archaeologists found that he had been buried in his clothes, a great honour, which indicates that he was considered a martyr. The injuries inflicted on him, clear evidence of execution, were described:

> The third cervical vertebra was severed by a sharp instrument in such a way that the main portion of the body and an arc of that vertebra were cut cleanly; the blow, struck from the left, also cut through the right corner of the lower jaw and its lower edge.

The observatory was destroyed soon after his death. His brutal son failed to keep the empire intact, and his name is forgotten, but that of Ulugh Beg is immortalized by having several features on the moon named in his memory.

VIDI NIHIL PERMANERE SUB SOLE
I have seen nothing last for ever under the sun

5

RENAISSANCE I

The story of the sundial now reaches a gap, and we do not know for sure how it was crossed. Somehow sundial science returned to Europe from the world of Islam. Maybe it was as a result of the Crusades that this science was carried back. More probably it was the scientific knowledge of Arab Spain that moved northwards. The oldest surviving Islamic sundial was made in Cordova in *c*.1000. However it happened, it seems that by the fifteenth century the crude form of canonical dials, such as those at Chartres and Bewcastle, were outmoded and replaced by instruments that were based on sound scientific principles. One of the earliest known European medieval dials that is mathematically sound is the wall dial on the Jacobi church in Utrecht in the Netherlands, which bears the date 1463. It was nearly discarded on a rubbish heap during restoration of the church in the 1970s but was rescued by the director of a local museum who recognized its significance in the history of time measurement. A treatise on sundials was published in 1540 by the Flemish scholar, Gemma Frisius, who was a professor at the University of Louvain in modern Belgium. He was born in Friesland, a coastal province in northern Netherlands, which explains why he gave himself the name Frisius. His birth name was Regnier Gemma but he adopted a Latin version when he became a scholar. His parents were very poor and both died when he was still a young child. He had a rough start in life, and had to overcome both poverty and physical disability caused by severe weakness in his legs, which could scarcely support his body. Though not as famous as his illustrious pupil Mercator, maker of maps and globes, Gemma became one of the leading scientists of his age. His university was renowned at that time for its mathematical and astronomical studies and there were close contacts between its scholars and those in Spain who had inherited the Arab learning. What is certain is that when the science of sundials came to late medieval Europe it was in southern Germany and the Netherlands that it first took root.

Most people imagine a sundial as a garden ornament, but today we tend to forget that people actually used sundials as common everyday objects; before clocks, they were the only way of

LEFT The sundial at the Jacobi church near Utrecht in the Netherlands is one of the oldest surviving European wall dials calculated on the principles we are familiar with today. It is dated 1463, is inscribed in Gothic script and tells equal hours from a slanting gnomon.

ABOVE Universal astronomical ring dial, Louvain, c.1550, unsigned, diameter 6⅛ inches. This is an example of the rare Renaissance instrument, beautiful in its simplicity. These dials were developed from the armillary sphere and from them was derived the universal equatorial ring dial, described later.

RIGHT Gemma Frisius is depicted in his study with a celestial sphere in his hands – the fish of Pisces can just be seen. Hanging behind him and to his side are an armilliary sphere, a quadrant (quarter circle) and an astrolabe. On the table are his writing and drawing instruments and a universal astronomical ring dial.

telling time. As watches are to clocks so hand-held dials were to outdoor ones. These pocket instruments, which had become popular by the seventeenth century, were often provided with a compass so that they could be pointed correctly. The invention of the clock did nothing to diminish the importance of sundials. Early clocks were hopelessly inaccurate and the worst offenders had to be reset every few hours. Sundials told time in unequal hours, but also in equal hours like the clock. It is somewhat strange that the invention of the clock stimulated the manufacture of dials needed to reset it. This disparity persisted for several hundred years, and explains why so many portable dials survive in museums and private collections.

The main centres of manufacture were Nuremberg and Augsburg, and to an extent London. Many of the dials that were produced were compendia – that is, gadgets that did several things, much like the mobile telephones of today. Many were very richly engraved and were commissioned as the presentation gifts of monarchs and grandees. These developments were taking place at the same time as the maritime exploration of the world was beginning and the mysteries of planetary motion being unravelled. It was fashionable for the elite to study or at the very least pretend to understand navigation and basic astronomy. Not all these dials were sumptuous, but even the simpler ones were engraved and decorated because they were always owned by people of means.

In Nuremberg the artisans were called compass makers for this had been their manufacturing trade for many years. Their speciality was the ivory 'diptych dial'. It consisted of two leaves that folded together, and formed a right angle when open. Ivory was a luxury material, easy to paint, and showed the sun's shadow very well, and the availability of African ivory increased at this time. The trade route had been overland across the Sahara and thence by Venetian or Genoese ship across the Mediterranean. As the Portuguese explored the coastline of Africa they took this trade and imported ivory in vastly bigger quantities. The ivory was painted with many pigments, burnt ivory itself for black, verdigris (which means 'green of the Greeks') made by exposing copper to the fumes of heated vinegar, blue out of cobalt oxide which first came from Bohemia, and orange/brown from the tree *Caesalpina* found in South America. These colours added great beauty to the engravings on the dials, and made it much easier to distinguish the many different calibrations.

Augsberg specialized in brass, and produced even more complex compendia. The fame of the city as a source of scientific instruments rested mainly on the work of two men, Christopher Schissler and his son Hans Christoph. The father was born in 1531 and is regarded as one of the greatest instrument makers in the Renaissance. He worked in the finest of metals, normally gilt brass, but occasionally silver and his work was clearly intended for and sought after by the very wealthy. His surviving works include sundials, compendia, astrolabes, quadrants, armillary spheres, globes, dividers and compasses.

At the same time the English were beginning to make fine instruments, a tradition of science and craftsmanship that would persist into the clock- and instrument-making skills of the

LEFT AND BELOW Three ivory diptych dials from Nuremberg. All three are by masters famous in their time – A by Paul Reinman, 1607; B by Thomas Tucher, undated; C by Leonhart Miller, 1638. They each have a compass for orientation and fold into a pocket-sized compact. Together they illustrate the different features of the diallist's science and art – equal hours at different latitudes, Italian hours, Babylonian hours, unequal (temporal) hours, dials to tell the zodiac, a scaphe dial and tables of latitudes. Each one is about 4½ inches long.

C

Christopher Schissler made this gilt brass astronomical compendium in Augsburg in 1563. It is pocket-sized, only 2½ inches long, most complex, and is a very fine example of his work. It would have been of particular use to a traveller and would have worked from northern Italy into Germany well to the north of Augsburg. Clearly it was the property of a wealthy person. It is a compact which folds out into three leaves, each of which is engraved on both sides to give six faces. These are exquisite in their detail and the spandrels and other unused spaces are richly decorated with relief work, produced in wax followed by etching. In the main photograph two other items can be seen which belong to the

instrument and accompanied it in its box. The wind vane's support rod plugs into the central hole on the furthest plate; this plate can be folded flat and the thin and light metal flag rotates on its rod to show the wind direction. A small folding tripod, which originally had a tiny weight hanging by a thread from the top, provides a plumb line for levelling the instrument.

SOL ME PROBAT UNUS

Only the sun can prove that I am useful

PLATE 1 Nocturnal dial with volvelle (a disc which revolves) and day and night scales. This is a device for time telling by means of the stars at night. To use it the date must be set first. The small tab can be seen in the illustration set against mid-January (IANVARIVS). The instrument is held vertically up to near the eye and the pole star sighted through the small hole in the centre. The rotatable arm (marked REGVLA NOCTIS) is turned until it is in line with two stars of the Plough constellation, Dubhe and Merak, which are known as the guards

of the Plough, because they lie on a line which points to the Pole Star. The time is then read on the second ring from the outside on which the hours are marked in Arabic numerals from 3pm to 9am. In the photograph the arm reads 5am. There are also knobs at each hour to assist hour reading in darkness by their feel. There is a further feature: inside the hour ring are three further rings which give the lengths of day and night in three latitudes, 51°, 48° and 45° (explained QVANTITAS DIEI, etc.). The scales are in cutouts which show

the values engraved on the disc underneath.

PLATE 2 Wind ring and lunar volvelle. There is a wind ring marking twelve compass directions in German. The circular aperture near the centre is a pictorial display of the phases of the moon. This is set by knowing and setting in a small cutout the age of the moon (the number of days since new moon). Knowing south from the compass on Plate 5, the time is read when the moon is seen passing south by aligning the instrument with the compass directions. In the centre

are four engraved boxes. These are collectively known to astrologers as an aspectarium because they give the different angles (or aspects). The geometrical engravings – a triangle (trine), a square (quartile) and a hexagon (sextile) – were of interest to astrologers because they refer to the different angles between celestial bodies which are astrologically significant; engraved here they are however merely decorative. The wind vane can be inserted into the centre so that the wind direction may be judged.

PLATE 3 Horizontal sundial. The gnomon is folded flat when not in use. It is adjustable for use in four different latitudes: 45°, 48°, 51° and 54°. There are four rectangular scales on the dial (chapter rings), one for each of the four latitudes. There is a cut-out hole through which the compass, can be read mounted on Plate 5, for aligning the dial.

PLATE 4 Latitude table in two columns for thirty-two German and Flemish cities.

PLATE 5 Compass. There is a degree circle and the inner ring is marked with the four cardinal points in Latin (SEPTENTRIO, ORIENS, MERIDIES, OCCIDENS). The compass is set on a rotatable disc and can be adjusted for magnetic variation.

PLATE 6 Stereographic projection through the tropic of Capricorn. This is used to convert between equal and unequal hours for places in a latitude of 51°. Put simply, the earth is sliced through a point on this tropic which is shown by the outer circle. It is also similarly sliced through the equator and the tropic of Cancer which are shown by the other two circles. The curved lines are unequal hour curves, projected on to the Capricorn plane and on the outer circles are marked equal hour times. A declination scale is marked on the XII line, according to the twelve zodiac signs. A thin thread is stretched from the centre hole to a point along an unequal hour curve, the point being also on an imaginary arc corresponding with the declination of the sun for that day. Where this thread crosses the outer rim of equal hours shows the time converted into equal hours.

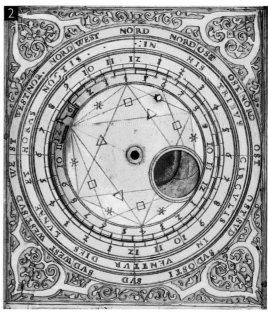

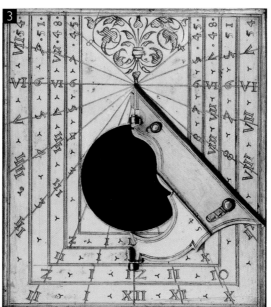

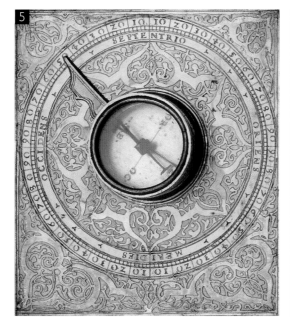

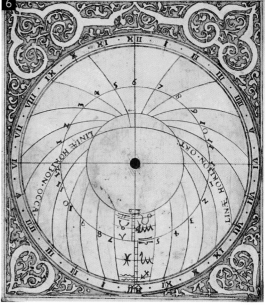

eighteenth century for which London became world pre-eminent and famous. Humphrey Cole was the greatest of these early English makers and is renowned for many fine instruments, which included armillary spheres, astrolabes, astronomical compendia, nocturnals, quadrants, rules and sundials. He supplied the instruments for Martin Frobisher's expedition in 1576, which attempted to find a north-west passage, and which led to the bay in Canada being named after him.

By the time the science returned to Europe, trigonometry was not much advanced, because the logarithms needed to make it really useful were not invented until the early seventeenth century. So dials were largely designed using the tools of a modern school geometry box – that is, pencil, ruler, compasses and protractor. Manufacturers were as usual under pressure to cut costs, and scales were made to reduce the drawing time, tables prepared and occasionally some rather complicated machines devised which eased the setting out of the lines and curves. The manufacture of these instruments and beautiful brass rulers and scales stimulated even further the skills of Europe's instrument industry. For a modern maker of sundials the best method is to use trigonometry, and the mathematics for horizontal and vertical dials is explained in the Appendix (see page 128).

However, only the simplest reasoning is required to understand the equatorial dial, and it is from this dial that horizontal and vertical dials are derived. The geometry of the equatorial dial can be seen from a glance at the earth itself (see page 56). Imagine the earth as a hollow transparent sphere with the axis a solid rod from pole to pole. As it turns on its axis the sun will cast from the axis a shadow on to the surface of the sphere opposite from the sun. The shadow image of the axis will be dimly visible on the inside of the sphere. If the equator on this earth is divided into twenty-four equal divisions (for the earth rotates once in twenty-four hours), the point on the equator where the shadow is cast will give the time. This is the basis of the sundial we know today. It is the equatorial sundial, and from it all other sundials with slanted gnomons are derived. The rules are simple and as follows on page 59:

ABOVE Compendium, 1579, by Humphrey Cole. The various discs and rings all fold together into a circular compact, which was attached to a pocket chain. The device comprises an equatorial dial, magnetic compass, perpetual calendar and table of saints' days, table of latitudes of European towns, a nocturnal, and a table of the sun's passage through the zodiac.

ABOVE Butterfield compass sundial, late eighteenth century, by N. Bion, silver, width 3¼ inches. Craftsmen continued to make pocket dials into the nineteenth century. A well-known form, of which many thousands were made in different workshops, was called the Butterfield dial. Michael Butterfield, after whom they are named, was born in England but went to school in Paris. He became instrument maker to the King of France and had a working life in the years before and after 1700. He first developed the instrument, which is adjustable for latitude by increasing or decreasing the angle of the gnomon against the bird's beak, and then reading from one of the appropriate dials, of which there are three for each of three different latitudes.

BELOW A dialling scale, c.1700, by John Rowle, brass. An example of a modest but highly reliable rule made by an English craftsman at a time when England had started to manufacture the finest instruments in the world. To set out the design for a horizontal dial the length of a latitude value was measured with a pair of dividers and noted. This formed the base of a right-angled triangle, and the whole length of the hour scale the hypotenuse. This would have been drawn on paper. Time points were marked off on this hypotenuse and lines drawn to the right angle to make time lines. The design was then copied and inscribed on the metal to be engraved. These two scales and the others could also be used to set out vertical declining dials.

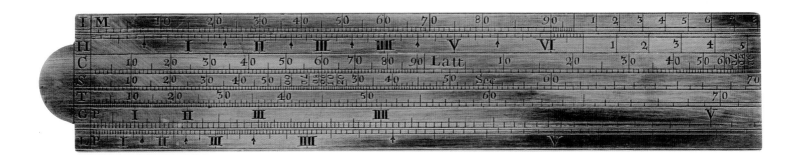

SOL IMMOBILIS TERRA VERSAT

The motion of the earth makes the sun rotate

ABOVE A view of the earth, the greatest sundial of all, from a point in space above the South Atlantic, taken on the Galileo mission launched in 1990. It is shortly before noon in London. The sun is low in the sky in the northern hemisphere, for whose inhabitants the season is at the depth of winter. There is sunshine nearly twenty-four hours a day south of the Antarctic Circle. Even so, the ice flows near the South Pole are extensive and the area is surrounded by a large blanket of cloud. A hurricane is developing between south-west Africa and Antarctica. The sun has just risen in Venezuela and will rise shortly in New York. It is late afternoon in India and night in Australia.

RIGHT The earth superimposed on to the great sundial sculpture by Henry Moore to demonstrate the three fundamental rules (see page 59). The dial is shown below in its original location in Printing House Square, before it was moved to the European headquarters of IBM in Brussels.

OPPOSITE Equatorial sundial, by Joanna Migdal, phosphor bronze, diameter 40 inches. This sundial was commissioned by the National Weights and Measures Laboratory. On the band are engraved words taken from Magna Carta, using the same lettering as in the original document, now in the British Library. The Magna Carta provided the first national definition of standards in England and gave birth to what is now this laboratory. The dial is mounted on a bronze turned weight, which is the laboratory's logo.

1. The gnomon must be at such an angle that its edge (or the centre of the rod), known as the style, is parallel with the earth's axis. As will be demonstrated at the start of the Appendix, it follows that the angle of the style with horizontal will be equal to the latitude of the place.

2. The ring that carries the numerals to mark the hours must be centred on the slanting edge of the gnomon and its plane perpendicular to it; in other words the ring must be parallel with the equator.

3. The numerals themselves must be spaced at intervals of $^1\!/_{24}$ of this circle of 360°, in other words at 15° ($24 \times 15° = 360°$).

There is one question that should be asked and answered: if the earth itself is a sundial rotating on its axis, the sundial is not rotating on the same axis as the earth's, for it is not at the centre of the earth, but on its surface. The two axes are different, and seem to us with our earthbound minds to be far apart. So how does the dial tell the right time? The answer is this: if the sun were the size of a football the earth would be no bigger than a grain of rice, and some 80 feet distant. So we can ignore the problem, which is insignificant, and assume that the sundial behaves as if it were at the centre of the earth.

There is a further matter of importance. Hilaire Belloc was witty but wrong when he wrote: 'I am a sundial and I make a botch/Of what is done much better by a watch.' The trouble with sundials is that they do not tell the correct time. That is certainly what most people think, but is it true? You could just as easily say that sundials tell the right time but clocks do not. It all depends on how you measure time. We have already seen how we once used hours that were longer in summer than in winter. We later began to use hours of equal length. But there have been other differences as well. The clock was invented at about the same time that Arab science brought sundials into Europe, but the early clocks were hopelessly inaccurate. The sundial did not suffer a decline with the introduction of clocks, for dials were still very much needed to reset them, and early clocks were sometimes out by at least an hour, sometimes by as much as two. Sundials are often to be seen next to clocks on church steeples. Man continued to live by the sun, not the clock. Gradually clocks became more accurate and clock hands moved at a constant speed. We have Galileo to thank for discovering the properties of the pendulum in the early seventeenth century, one of the most useful inventions for mankind, which remained of prime importance for nearly four hundred years. Pendulum clocks are now obsolete and have been replaced by electronic timepieces but their beauty remains a delight. Clocks move at a constant speed – they are 'isochronous', but does time move at a constant speed? Most people would answer yes without hesitation and think the question absurd, but after a moment's thought one must accept that it is impossible to know that time moves at a constant speed. It is only because we live by the clock that we make the easy assumption about the even flow of time. But the sun certainly does not move at a constant speed across the sky or around the earth. So there is usually a discrepancy between sun time and clock time. When we say 12 o'clock we are using language which goes back at least as far as Chaucer, who used the phrase 'of the clokke' to distinguish between time by the clock and time by the sun. Chaucer was himself an accomplished astronomer and author of a *Tretise of the Astrelabie* (*c*.1391), possibly the first scientific treatise written in the English language.

To astronomers, sun time in any place is called Apparent Solar Time, and clock time in that place is called Local Mean Time. Note the word 'mean', which surely suggests that while the dial is accurate, the clock is right only on average. The difference between sun time and clock time varies every day of the year, but the values are pretty much the same from one year to the next. The dial can be up to fifteen minutes ahead of or behind the clock. Astronomers call this difference the Equation of Time (EoT). While there is usually a difference between sun time and clock time, a glance at the illustration below will show that there are four moments of the year when this is not so: in April, June, September and December, when the EoT is zero.

In 1609 Johannes Kepler published his profound and brilliant conclusions about the earth's orbit round the sun. Two of his discoveries are relevant to this subject. Like all the other planets, the earth circumnambulates the sun in an elliptical orbit, which means that its distance from the sun varies throughout the year; Kepler also proved that when the earth is further from the sun it moves in its orbit more slowly than when it is closer. The axis of the earth is also tilted against the plane of its orbit, which accounts for the change in the seasons. These two factors – the varying orbital speed of the earth and the tilt in its axis – cause the Equation of Time. The illustration of the seasons on page 27 demonstrates the orbit. Kepler proved that the earth moves round the sun more slowly in midsummer and midwinter than it does at the equinoxes. The illustration also shows the tilt of the northern hemisphere towards the sun in summer and away from it in winter. The table shows graphically how the two differences combine to give values for the Equation of Time.

There is yet another factor that has undermined the reputation of sundials for accuracy. In Britain we do not live by local mean time; we live by the time at Greenwich, except of course during summer time when we add an hour. Other countries have different time zones. The British time system is still popularly called Greenwich Mean Time (GMT) because it is based on the Greenwich meridian. There is about 17 minutes' difference between sun time at Greenwich and at Plymouth.

It follows that in addition to the adjustment for the Equation of Time you must also make an adjustment by adding to or subtracting some minutes from the dial readings to allow for the

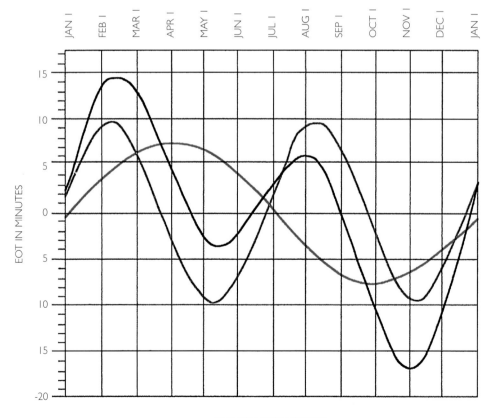

THE EQUATION OF TIME

A table showing the difference between the time of the sun and the time of the clock. The effect of the earth's elliptical orbit is shown by the red curve, and the difference caused by its tilt is shown in green. The Equation of Time is due to the combination of these two separate effects and is shown in black.

difference in sun time between the national or regional time meridian and the dial's location. The sun rises in the east and, for every degree of longitude west from the meridian, you must add 4 minutes to dial times, and eastwards you subtract it. The reasoning is this: there are 360° of longitude round the earth and 24 hours in a day, and 4 × 360 minutes = 24 hours. These two corrections will henceforth be referred to as Equation of Time and displacement.

All this makes for complication – particularly if in addition you must add a further hour for summer time, but, in the sundial's heyday, the time meridian was for most people where they were. Only gradually did Greenwich Mean Time come into use. It was referred to in railway timetables and was popularly known as railway time. Until early in the twentieth century, every station manager in France knew how to read a sundial and correct its times to the Paris meridian. As late as 1858 in Britain there was a legal case in which the outcome turned on whether the case commenced at 10am Greenwich time or 10am local time. Even though almost all British public clocks were by then set to GMT, legal arguments in favour of local time prevailed, and it was not until 1880 that the British Parliament legislated to make Greenwich Mean Time the legal standard for the country.

The two corrections – Equation of Time and displacement – can be combined into a table for use in reading the dial. It is also possible to incorporate them into the design of the sundial, but this is not completely straightforward. One solution is the noon dial, an instrument that only indicates noon, but corrected for the Equation of Time and displacement. The Radcliffe Observatory dial shows how this results in an analemma, a curve that follows a figure of eight. You can see on this dial where the sun's image will be projected at noon GMT (or 1pm BST) at any date of the year. Of course this only tells the right time once in a day, but when sundials had the practical use to set clocks, once every day that the sun shone probably sufficed. Doug Bateman has designed a noon mark for viewing from inside a window. The glass was shot-blast with the analemma, and the spot of light is projected on to it from outside. It is probably the most accurate wall dial in the world for direct readings of noon.

In order to show *all* the times of the day with clarity and corrected it is necessary to have two dials, one for when the days are getting longer and another for when they are becoming shorter. This is because the sun rises to the same height at the same hour twice in the year but the Equation of Time is different on those two dates. For example, at the spring equinox (20 March) the EoT is about +7 minutes; at the autumnal equinox (23 September) the EoT is about –7 minutes. The vertical dial in marble illustrated on page 64 shows a solution. This has four dials. One pair faces south-west; the other pair south-east. An even more sophisticated solution is found in the Greenwich sundial. This is an equatorial dial with corrections which are incorporated in the engravings. There are two metal plates, which are changed round at the time of the solstices. The time is read from the centre of the gap in the shadow made by the two dolphin tails.

When Europe had relearned the science it had earlier forgotten, it still lived by the sun, not by the clock. Furthermore, time was the time of a particular place, not a particular time zone, and the notions just described were irrelevant and still unanticipated.

UTERE, NON NUMERA
Use them, don't count them

MERIDIES MEDIA
Mean noon

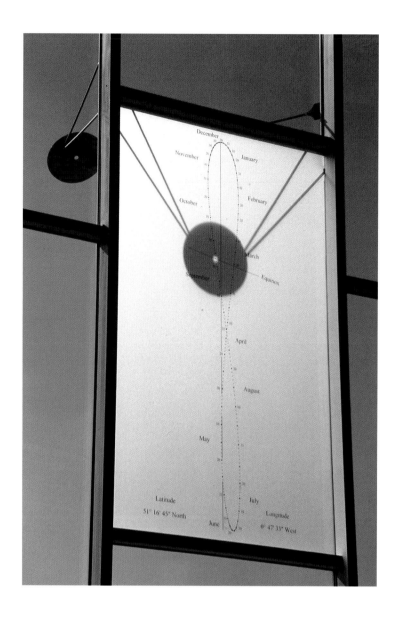

LEFT This fine noon mark was designed by Christopher St J.H. Daniel and Martin Jennings to commemorate the bicentenary of the Radcliffe Observatory in Oxford. An image of the sun is projected through a hole in the centre of the sun symbol at the top left. The sun will move along the curve at the bottom in midsummer and the curve at the top in midwinter. At the equinox it will follow the straight line. The analemma – the figure of eight in gold – has been calculated to allow for the difference between sun time and clock time at any date.

ABOVE Glass sundial noon mark, 1996, 6 feet 7 inches × 3 feet 3 inches. Doug Bateman constructed this large and accurate noon mark for the Cody Building at the British Defence Research Agency, now known as QinetiQ Ltd, Farnborough, UK.

RIGHT Dolphin equatorial dial, National Maritime Museum, Greenwich, by Christopher St J.H. Daniel and sculpted by Edwin Russell, bronze on a plinth of Portland stone, dial plate width 25 inches, height 14 inches. This dial was commissioned to mark the Silver Jubilee of Queen Elizabeth II.

OPPOSITE Rosemoor dial, Royal Horticultural Society, Devon, 2004, height *c.*7 feet. This dial was made by the artist, stonemason and carver Ben Jones to a design by Mark Lennox-Boyd. The top two faces are used when the days are getting longer and the bottom two for the other half of the year. The time lines are curved and the exact time is read from the tip of the pointer on the appropriate face, because the curves have been calculated to allow for the Equation of Time and the displacement from the Greenwich meridian. It stands in the rose garden, and the inscription on the base is from *Four Quartets* by T.S. Eliot, and reads: 'But only in time can the moment in the rose garden be remembered.'

RIGHT Design for engravings of a vertical dial in marble, 1987, by Mark Lennox-Boyd. These engravings, made for a site in the north of England, clearly show the idea behind the Rosemoor dial.

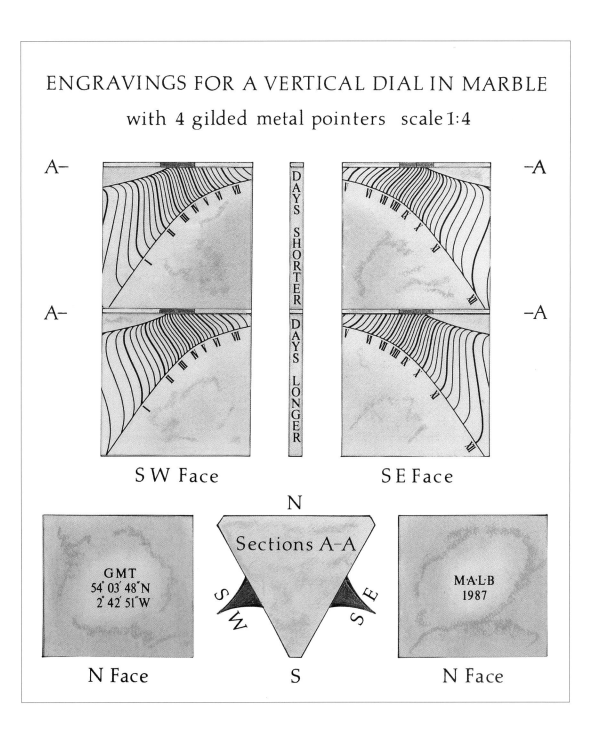

ENGRAVINGS FOR A VERTICAL DIAL IN MARBLE
with 4 gilded metal pointers   scale 1:4

A–   –A

DAYS SHORTER DAYS LONGER

A–   –A

S W Face   S E Face

N

Sections A–A

GMT
54° 03′ 48″ N
2° 42′ 51″ W

M·A·L·B
1987

N Face   S   N Face

6

RENAISSANCE II

When Europeans rediscovered the sundial they learnt for the first time about the slanting gnomon and firmly established the principles that underlie the dials we see today on pedestals and walls. Yet the great majority of dials manufactured in the sixteenth and seventeenth centuries were not designed for open spaces but for carrying on the person, and this extensive industry continued until the development of accurate pocket watches. Renaissance men were also inspired to develop beautiful and original ideas for sundials in other ways. These ideas would again inspire others, but it was only at the end of this spectacular period that a need to live by the time of the clock rather than the sun became widely accepted. In the sixteenth century, scientists knew that the earth's axis was tilted, but they still did not know that the earth's orbit and those of the other planets were elliptical. It was not until the early 1600s that the ancient Greek belief that the planets moved in perfect circles was finally disproved as a result of the work of Tycho Brahe (1546–1601) and Johannes Kepler (1571–1630). It was the latter who cleared the path to a full understanding of the Equation of Time.

Rarely in the history of science have two more dissimilar people been thrown into companionship. Tycho Brahe was a great Dane – aristocratic and arrogant. Born with wealth and every possible connection he entertained the great and good of Europe with gargantuan meals in his eccentric island observatory palace named Uraniborg after Urania, the muse of astronomy in Greek mythology. He housed a dwarf jester and a pet elk, which died after raiding a beer store, and threw his servants into the palace jail for even minor transgressions. In his youth he had lost his nose in a duel and, ever after, strapped a false nose to his face. This was made of electrum, an alloy of gold and silver, one might say an appropriate material, for the metal was also used by the ancient Egyptians to cap their pyramids so that they would reflect the earliest of the morning's sun, just before daybreak. Kepler, by contrast, was a pauper who came from a family of misfits. The little cottage in which he was born in Weil, Swabia, in south-west Germany, had few rooms for the family of more than twelve. They lived upstairs and were kept warm in winter by the

RCIS VRANIBVRGI, A TYCHONE BRAHE, DÑO DE KNVDST

NSVLA HELLESPONTI DANICI HVENNA CONSTRVCTÆ, QUO AD TOTAM CAPACITATEM, DESIGNAT

OPPOSITE Tycho Brahe and Johannes Kepler, two remarkable men who quarrelled but complemented each other. Brahe provided the accurate data, accumulated over many years, Kepler the interpretation and calculation, developing one of the most brilliant deductions ever seen by mankind.

ABOVE Uraniborg. On the island of Hven, in Denmark near Copenhagen, Tycho Brahe constructed his extraordinary castle observatory. Coloured engraving by Wilhem Blaeu from *Nouvel Atlas* (Amsterdam, 1645). The garden had images of the stars in its topiary and, to the left of the observatory, a sundial can be seen.

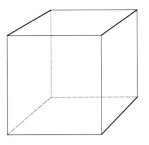 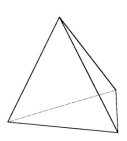 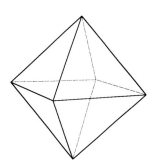 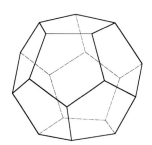 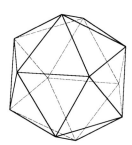

The Platonic solids:

• Cube, six squares
• Tetrahedron, four equilateral triangles.
• Octahedron, eight equilateral triangles.
• Dodecahedron, twelve pentagons.
• Icosohedron, twenty equilateral triangles.

These shapes had a profound influence on the minds of Renaissance scientists. The great Kepler constructed designs in which each is inserted into a sphere which touches the apexes – a sphere which in turn can be inserted into the next shape, touching its faces, until all five, with their spheres, make a model, one solid inside the next one. He then erroneously reasoned that the diameters of the spheres were in proportion to the distances of the major planets from the sun. No one knew these distances precisely but he was driven by the belief that, because the universe was divinely created, it followed that it must display the beauty of harmony and symmetry. Such is the power of an imagination that seeks to fit uncertain facts to false theory.

farm animals in the rooms below. His great manuscript is lovingly preserved and treated like a holy relic. He is the town's hero and venerated like a patron saint. Yet his childhood was intolerable and all his life he suffered from ill health and emotional deprivation. His father narrowly escaped the gallows for desertion and his mother the stake, for witchcraft. But the insecure youth had a good Lutheran education provided by the Dukes of Württemberg, who had created a modern educational system. He was precociously brilliant.

The two men, though both irascible and frequently quarrelsome with each other, were united by a passion for astronomy. They were opposites, who loved each other one day and filled the next with resentment. For more than twenty years before they met, Brahe and his assistants had mapped and tabulated the heavens and the movements of the planets to an accuracy never previously even imagined. The telescope had not been invented and they did so with crude instruments and good eyesight, repeating and averaging their observations. Kepler, the genius of the two, used these tables to understand the eccentric and inexplicable movements of the planet Mars. Why did it normally move, like the moon, in one direction, and then apparently reverse itself? Why did it not seem to move at a constant speed? He had been invited to join the court of Brahe, who assigned to him the task of understanding its mysterious movements, and, with Brahe's data, he had the information on which to work. Brahe, poor man, died in great pain from a blockage in his bladder, and during his last night of delirium was heard to implore God with repeated murmuring: 'Let me not seem to have died in vain.'

He did not, for without his tables Kepler could have made no progress. He had at first thought his task would be easy, and accomplished in several weeks, but he was much mistaken. For six long years he laboured by daylight and by candle at the most dreadful mathematics, asserting and demolishing, by painstaking calculation, one false hypothesis after another. He made many mistakes, pursued false paths, had terrible disappointments and, when he discovered his errors, was forced to retreat, diverge and continue, and then find yet another error and start once again. He made many hypotheses about how the planet moved and suffered much torment. At one point in his great book *Astronomia Nova* (A New Astronomy), published in 1609, he wrote: 'If thou [dear reader] art bored with this wearisome method of calculation, take pity on me who had to go through with at least seventy repetitions of it, at very great loss of time.' You can be sure that each of these calculations took many hours. For one ghastly year, during which he had no money, he tried to prove that the orbit was egg-shaped, finally recording, rather imaginatively, that his egg 'had gone up in smoke'. Often the truth was near his grasp but he failed to snatch it, and he then deviated down the wrong side way. He was driven by a will of iron. He was absolutely certain that there had to be a clear rule underlying the planet's movements, because they were ordained by God, and it was this that made him persevere in this dreadful endeavour. He continued with his struggle until one day, in a moment that must have been the most charged with emotion in his whole life, he knew he was right. After 900 closely written and smudged folio pages of terrible calculation, false paths and painful setbacks, he triumphed. He had proved that the orbit of Mars was an ellipse.

It was not much of an ellipse, in fact very nearly a circle. If one were to draw the orbit accurately, but vastly reduced in scale, on a page of this book, one would not notice the difference. Kepler's proof was an astonishing achievement. Arthur Koestler has written: 'Although . . . a person may never have heard of Kepler's laws, his thinking is moulded by them without his knowledge; they are the invisible foundation for a whole edifice of thought.' Kepler's work had an impact on educated people as great as the cosmological discoveries of today. It was soon demonstrated that the orbit of all planets was elliptical, and the way was open for scientists to understand and calculate the Equation of Time.

When Renaissance scientists rediscovered the sundial, they were fascinated by other aspects of Greek mathematical learning that had come at the same time, for example the Platonic solids, named after the ancient philosopher. There were thought to be only five regular solids, that is, solids with identical faces and identical angles at each vertex. In fact there are four more, vastly more complex, but beautiful; two were discovered by the great Kepler, two others not until about 1800. The various types of solid have exercised a fascination over the minds of mathematicians of all ages, among them some of the greatest names in mathematics. In Greek times and to some Renaissance men they were believed to hold mystical properties. After all, the followers of Pythagoras had considered mathematics a religion and many Renaissance men believed that geometry was divine knowledge. Kepler himself wrote:

> Geometry existed before the Creation, is co-eternal with the mind of God, is God himself . . . geometry provided God with a model for the Creation and was implanted into man, together with God's own likeness – and not merely conveyed to his mind through the eyes.

By the end of the sixteenth century, sundial makers decided to put dials on these solids. Some were richly engraved in gilt brass, some, more modestly, were made of paper glued to wood. A sundial can be made on any surface reached by the sun. The dial projections are determined by the geometry, occasionally horizontal, occasionally vertical, but sometimes neither. These latter are called reclining dials, which recline from the vertical but still face south. When they also do not face directly south they are called reclining declining dials. They were designed no doubt to show the skills of the craftsman-mathematician who devised them, and the beauty of their shape. Other wonderful geometrical shapes that were not Platonic solids were also used, like the

TOP RIGHT Polychrome decorated polyhedral dial, Italian, nineteenth century, unsigned, wood, height 9 inches. The polyhedron is identical with the dial on the left but only seventeen of the faces are furnished with dials and it is attached to its support in a different manner. It is painted in gilt with floral reserves in red, green, blue and brown, with iron pin gnomons, on a baluster turned support and shaped foot.

FAR RIGHT Polyhedral dial, 1587, by Stefano Buonsignori, wood, height 7½ inches. The instrument is in the shape of a regular dodecahedron and finely decorated with brilliant colours. Each face is engraved with a different form of sundial – scaphe and reclining declining. The hollow space at the top was for a magnetic compass (now missing) for orientation. One face bears the Medici coat of arms. The maker can be deduced from the initials 'D.S.F.F.', which stand for 'Don Stephanus Florentinus [or Florentiae] Fecit'.

BELOW RIGHT Polyhedral dial, sixteenth century, by Stefano Buonsignori (attr.), wood, height 6¼ inches. It has triangular faces in the form of an octahedron, decorated in the typical style of Stefano Buonsignori. It is set for the latitude of Florence and is fitted with a magnetic compass for orientation. Each face is painted with a different type of sundial – horizontal, vertical and reclining declining.

LEFT Polyhedral dial, Munich, 1578, from the workshop of Hans Koch, silver gilt, height c.14 inches. This is an exceptionally fine example of the kind of work that was very costly and often undertaken as a presentation gift to a person of great eminence or political importance. There are twenty-five dials with pin gnomons on a shape known by the rather bewildering name of 'small rhombicuboctahedron'. It is furnished with a compass at the top for orientation. The face marked SEPTENTRIO points north.

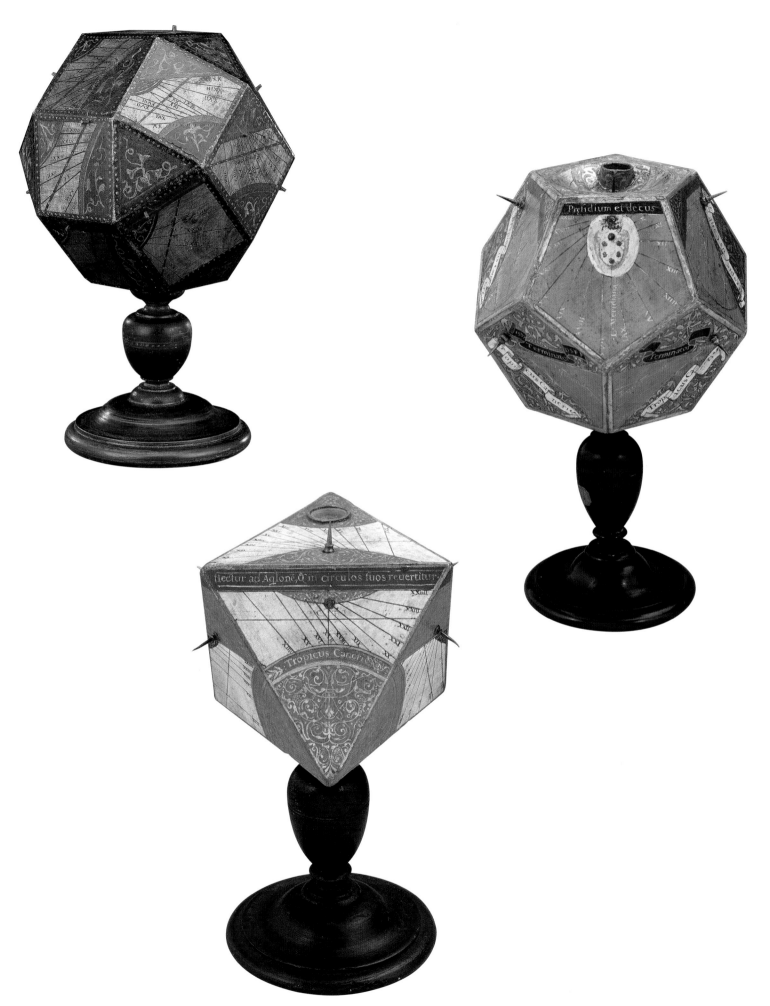

BELOW Gold finger-sized ring dial, German, sixteenth century, probably from Augsburg, diameter 1 inch. It is engraved on the outside with the initials of the months, on the inside with an hour scale 4–12–8. It has a collet-set single cabochon malachite. Finger ring altitude dials of this type are rare, and this example was probably made for a wealthy client, since malachite was an expensive stone at the time. Its use also suggests that the dial was made in the Augsburg area where malachite was mined.

In Shakespeare's *As You Like It*, Jacques says of a fool he met in the forest: 'And then he drew a dial from his poke, And looking on it with lack-lustre eye, Says very wisely, "It is 10'clock".' (Act II, scene 7). The dial was almost certainly a ring, probably less richly decorated than the illustrated example, which was kept on a pocket chain from which it was suspended for reading. The dial works from the sun's altitude, the hour is roughly judged from an image projected through a small hole in the opposite side of the ring. The lug at the top was made to slide round the ring to adjust the dial for date; the months can be seen engraved on the outside.

It was recorded that, at his execution, Charles I had either a ring dial or a circular slide rule in his pocket. It seems strange that on such a day he took either object on his person, but surely it is more likely that it was a ring dial.

ABOVE Chalice dial, British Museum, silver gilt, height 5½ inches. A chalice dial is another form of altitude dial. The readings are marked inside the conical cup which has a gnomon centrally placed and reaching the top. The scales are then calculated for the inverted cone, which forms the inside. As can be seen from the photograph, it was made in 1554 by Bartholomew, who was the Abbot of Aldersbach, a small town on the German-Austrian border. No doubt it was used for communion wine but must have been rather awkward for anyone sipping from it. This cup is the earliest known surviving example.

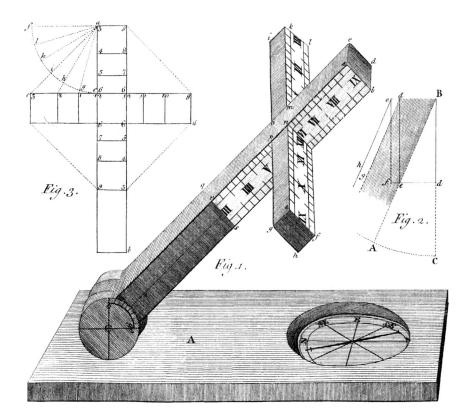

ABOVE Cross dial, from *Lectures on Select Subjects* by James Ferguson (London, 1773). The dial points south and the time is read from shadows cast by the edges of the arms and top on to the adjacent arm or vertical. For obvious reasons they are sometimes found in churchyards.

ABOVE RIGHT Stained glass window, Town Hall, Ulm, 1560, height 24 inches. Sundials on walls must be large if they are to be legible from the ground, but they can be small and exquisite if they are depicted in stained glass on windows and read from the inside. Many are to be found in European buildings. The dial at Ulm (the birthplace of Albert Einstein) depicts a master and his servant, and contains a message of human mortality and equality.

'The Lord will die and equally the servant, the good as well as the evil, and nobody will know in the morning, whether he is still alive in the evening.'

magnificent example shown on pages 70–1, a shape referred to by a grand phrase 'small rhombicuboctahedron'.

Dials were made on these solids, called polyhedra, and also on spheres, on crosses for the pious, on stained glass windows, pillar dials, chalice dials, on rings or on other everyday objects. They were made by reflecting a spot of light from a mirror on to a ceiling. Art and science were combined with ingenuity to explore new areas of design.

It is recorded that Thomas Jefferson made a spherical dial, but it is lost. A reproduction has been made for the grounds of his estate at Monticello in Virginia. Although the idea behind the design was well known in Europe it is possible that the President reinvented it because in August 1816 he wrote to his architect friend Benjamin Latrobe:

It occurred then [to me] that this globe might be made to perform the functions of a dial. I ascertained on it two poles, delineated its equator and tropics, described meridians at every 15 degrees from tropic to tropic, and shorter portions of meridian intermediately for the half hours, quarter hours, and every 5 minutes. I then mounted it on its neck, with its axis parallel to that of the earth by a hole bored in the nadir of our latitude, affixed a meridian of sheet iron, movable on its poles, and with its plane in that of a great circle, of course presenting its upper edge to the meridian of the heavens corresponding with that on the globe to which its lower edge pointed. . . . Perhaps indeed it is no novelty. It is one however to me.

Soon after 1400, perspective was developed by Filippo Brunelleschi. Its principles are familiar to the diallist, who projects his lines in a similar way. Some artists took an interest in the subject. In a book published in 1524 are some drawings of sundials by Albrecht Dürer. Father Bonfa spent a great part of his time between 1672 and 1673 delineating two sundials at the Jesuit seminary in

ABOVE Spherical dial, Monticello, USA. It is well known that the founding President of the United States had a strong interest in science and design. One of his creations was a sundial. This has been lost but a copy has been reproduced from surviving descriptions. It is a globe dial and is in effect a reproduction of the earth itself. The cursor is turned until its shadow is at its thinnest and the time is read. It sits on a pedestal of Jefferson's design and a cast-stone capital modelled on the 'corn cob' capitals designed by Benjamin Henry Latrobe for the old vestibule of the Senate wing in the US Capitol.

RIGHT Lycée Stendhal, Grenoble, France. In the seventeenth century the school was a Jesuit seminary. Following Jesuit tradition, Father Bonfa was greatly interested in astronomy, and aided by his pupils he depicted this extraordinary sundial which covers an area of more than 100 square yards. In the illustration the images of Gemini, Taurus and Capricorn can clearly be seen, together with the equinoctial line and the names of several Christian festivals in Latin.

The Palazzo Spada in Rome, from *Perspectiva Horaria* by Emmanuel Maignan (Rome, 1648). The Palazzo Spada is a magnificent Renaissance building dating originally from the sixteenth century but purchased in 1632 by Cardinal Bernadino Spada and much embellished by him. He commissioned the sundial which was designed by Father Emmanuel Maignan (who made another in Rome at the convent of the Minimes) and painted by Giovanni Magni in 1644.

It is in a barrel-vaulted room with windows down one side, in a central window of which is a horizontal mirror that reflects the sun image on to the vault. The dial tells equal, temporal, Babylonian and Italian hours, the sign of the zodiac, azimuth and altitude. The curves are in fresco, with different colours to distinguish the different information. The ceiling also indicates six of the planetary houses of astrology for the location, and some latitudes and longitudes of various places on the globe. In the room there is also a volvelle, for converting moon time to sun time, and an astrological table to show which planet rules the day of the week.

Grenoble. They are operated by two mirrors, which cast spots of light on to the ceilings of two flights of stairs, up and down which can now be heard the clatter of teenagers: the building has become a high school for girls. The greatest exponent of all was another Frenchman, Father Maignan, who made two magnificent dials in Rome, one at a convent and the other to adorn a barrel-vaulted room at the Palazzo Spada. In the United States Claude Hartman has depicted a sundial on an awning, inspired by these Renaissance precedents, a French Canadian couple on a public library window, a Dutchman on beer glasses. In the same tradition, a contemporary dial has been constructed by the author of this book in Italy at a house near Rome called La Meridiana. One inventive scientist, famous in his time, was Nicholas Kratzer, who came to England from Bavaria in about 1520. He had been educated at the universities of Cologne and Wittenburg. In England he was appointed as horologer and part-time diplomat to Henry VIII. He became tutor in Mathematics to the children of Sir Thomas More, lecturer in astronomy for Cardinal Wolsey at New College, Oxford, and a friend of Hans Holbein, who immortalized him in one of his most loved paintings. Another well known and popular painting is Holbein's depiction of *Jean de Dinteville and Georges de Selve*, known as *The Ambassadors*, which also contains images of sundials possibly as a reference to the passing of time.

There is a cylindrical dial in *The Ambassadors* and many fine ones in the museums of the world. As we know, the height of the sun varies with season and, if we know the date, the time of day can be determined by measuring the height of the sun. Such dials are more commonly known as shepherd's dials, since they were often the dials of farming people of modest means, and were still in use in the early twentieth century in rural Spain. In fact the dial was invented in Roman times and an early classical design dating from AD 100 was recently identified in the museum of Este in northern Italy. It had been excavated in 1884 but its purpose as a sundial was not identified for a hundred years. In 'The Shipman's Tale' Chaucer referred to one: 'For by my cylinder it is pryme of day,' reading, of course, in canonical hours. These dials operate by means of a rotatable cursor at the top. This is turned to the appropriate date and then the whole object is turned so that the cursor directly faces the sun and the shadow of its tip will read the altitude and the time.

Greek civilization established rules of science, which remained unchallenged until the Renaissance. Even the belief of the Catholic Church that the earth, not the sun, was at the centre had come from ancient Greece – a *doctrine* of Christianity based on erroneous ancient science. But when the rebirth of knowledge occurred in Europe, the most magnificent artefacts were produced. Science was integrated with art and design as never before in the West, and craftsmen manufactured instruments of great beauty and delight which until then had never been surpassed and are unequalled today. On the following pages are illustrations that depict Kratzer and his work, and other contemporary designs inspired by the inventions of Renaissance scientists.

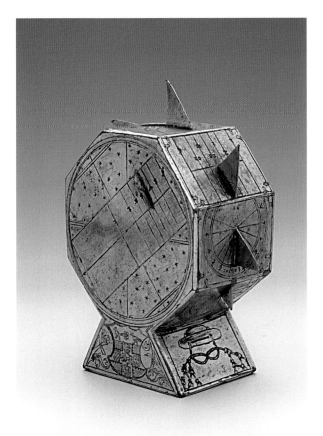

ABOVE LEFT Polyhedral sundial by Nicolaus Kratzer, 1518–30, c.4 inches high, gilt brass. This small dial is unsigned but it is almost certainly Kratzer's work as it has similarities with other dials by him and bears a cardinal's hat and the arms of Wolsey, for whom it was clearly made.

LEFT Shepherd's dial, Italian, late sixteenth century, maker unknown, wood, height 13¾ inches. This is a finely decorated cylindrical dial. On the surface of the column are drawn the hour lines, at the top of which are the months and the zodiac signs. The movable top section has two gnomons of different length, to be used according to the time of year. There is a painted dedication to Francesco I de' Medici, suggesting a date of manufacture between 1574 and 1587, the years when the eldest son of Cosimo I assumed the title of Grand Duke of Tuscany.

ABOVE *Nicholas Kratzer* (1528) by Hans Holbein the Younger. In England Kratzer soon became a good friend of Holbein. They were, after all, both of Germanic origin and moved in the same circles. This lovely picture hangs in the Louvre (there is a copy in the National Portrait Gallery, London) and shows the astronomer surrounded by his instruments and sundials. Holbein came from Augsberg and would surely have been aware of the beautiful sundials manufactured there. What is of interest is that many of the instruments in this portrait are identical to those in the Holbein painting *The Ambassadors*. He is holding a polyhedral dial, and on the shelf behind him are a semi-circular instrument with plumb line, and a shepherd's dial. This has led to the almost irresistible conclusion that Kratzer was Holbein's technical adviser for the picture of *The Ambassadors*.

BELOW *Jean de Dintevillle and Georges de Selve 'The Ambassadors'* (1553) by Hans Holbein the Younger. This famous painting in the National Gallery, London, includes the three sundials described in the text and various other astronomical instruments. Jean de Dinteville, on the left, sent as ambassador by the French King to the court of Henry VIII was unhappy in London, and in particular hated the weather. Scholars have long been puzzled by the complex imagery in the picture. Modern opinion holds that the astronomical and musical instruments are arranged perhaps to simulate heaven and earth and that the painting expresses thoughts of melancholy, despair on the religious divisions in Europe brought about by Martin Luther's schism, on a world of chaos, the brevity of life, and hope of the life to come.

RIGHT Beer glass dial, 2002, by Hendrik Hollander, Amsterdam. These glasses are corrected for the Equation of Time. Turn the glass until the spot of sunlight points to the correct date to see whether it is later than 5pm, the hour Mr Hollander recommends for filling his glasses with beer.

BELOW 'Et pourtant, elle tourne', Municipal Library, Bizard Island, Quebec, 1995, by Michèle Lapointe and René Rioux. 'And yet, it turns': this sundial is titled after a translation of the words of Galileo, allegedly whispered by him to recant his earlier recantation of the correct but impious assertion that the earth revolved around the sun. It is furnished with an external rod gnomon that casts its shadow on to the library window.

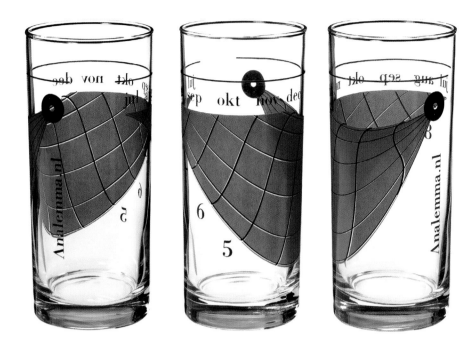

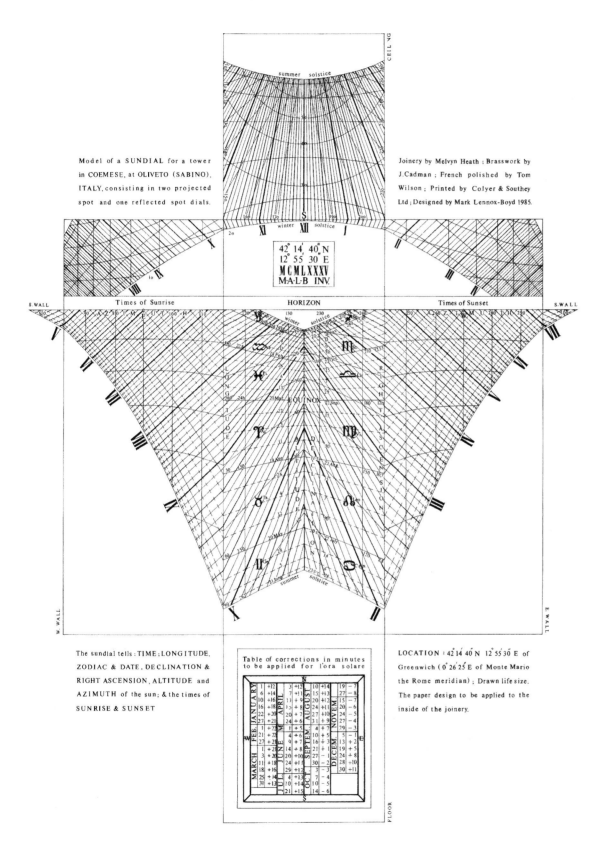

CEILING

summer solstice

Model of a SUNDIAL for a tower in COEMESE, at OLIVETO (SABINO), ITALY, consisting in two projected spot and one reflected spot dials.

Joinery by Melvyn Heath ; Brasswork by J.Cadman ; French polished by Tom Wilson ; Printed by Colyer & Southey Ltd ; Designed by Mark Lennox-Boyd 1985.

winter   XII   solstice

42° 14′ 40″ N
12° 55′ 30″ E
MCMLXXXV
M·A·L·B INV.

S.WALL   Times of Sunrise   HORIZON   Times of Sunset   S.WALL

W.WALL   E.WALL

The sundial tells : TIME ; LONGITUDE, ZODIAC & DATE , DECLINATION & RIGHT ASCENSION , ALTITUDE and AZIMUTH of the sun ; & the times of SUNRISE & SUNSET

Table of corrections in minutes to be applied for l'ora solare

LOCATION : 42° 14′ 40″ N 12° 55′ 30″ E of Greenwich (0° 26′ 25″ E of Monte Mario the Rome meridian) ; Drawn life size. The paper design to be applied to the inside of the joinery.

FLOOR

PAGES 80–3 La Meridiana, Oliveto, Italy, 1999–2005, by Mark Lennox-Boyd. La Meridiana is a house with a stair tower, inside which a large sundial is depicted. A background of skyscape in different moods of weather was painted by Dominique Lacloche in casein tempera and superimposed on this are depicted the lines and curves of the dial. It operates as a camera obscura. At dawn a spot of light is projected from an opening in the eastern wall, and moves across the walls opposite. Towards the middle of the day the sun reaches a horizontally placed mirror beneath the window on the south wall, and its image is projected on to the ceiling. In the afternoon the sun is projected from an opening in the western wall and moves across the opposite walls until sunset. As the sun sets its red tinge is projected by the camera obscura on to the opposite wall creating a warm red glow. At certain points the sun moves along the treads and upright edges of some of the steps, and in all there are twenty-four projections. It tells time to within a few seconds, date (in red) to within a day, altitude (blue), azimuth (green), the sign of the zodiac and the times of sunrise and sunset. The inscription is from Psalm 113: 'From the rising of the sun' on the west wall where the sun is seen to rise, 'Praise be the name of the Lord' on the north wall, 'Until its setting' on the east wall where the sun is seen to set.

The sundial occupied the builders, the artist, gilders, painters, metal workers, other craftsmen and the author more than 4,000 hours of work. There are certainly disagreements whether a number of records can be attributed to La Meridiana, but few would dispute that no one has ever before put more effort into one sundial.

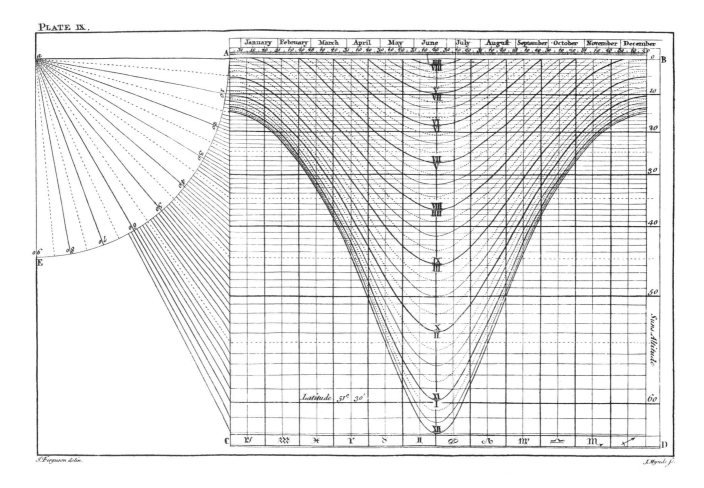

LEFT Design for a shepherd's dial, from *Lectures on Select Subjects* by James Ferguson (London, 1773). This design is to be wrapped round a cylinder. It is for the latitude of London, and most clearly demonstrates the features of an altitude dial. At the top is a scale of dates, at the bottom the corresponding zodiac divisions. The altitude of the sun is shown on the right hand side, and on the left, not part of the dial but an aid to its calibration, is a protractor showing solar altitude as well. The altitude of the sun at any time, for any date, in any latitude is given in many navigational publications, and the dial is plotted with this information. One can read from the dial, for example, that in London, on 19 April, at 8am or 4pm, the altitude of the sun will be 27°. Such is the quality of Ferguson's engraving.

LEFT BELOW Hartman skylight dial, 1999. Claude Hartman can relax in the shade of his porch at Arroyo Grande, California, under this dial depicted on the canvass of his awning. The dial tells the hour along each of the analemmas, marked by a slash of sunlight from a second awning, higher up.

RIGHT Shepherd's dial, 1992, by Ono Yukio, height 7 feet 10 inches. The humble shepherd's dial has been transformed into a modern sculpture. The gnomon can be turned to the sun and the hour curves are marked on a drum, which is rotated until the gnomon's shadow falls on to the correct vertical date line.

EASTER, THE CALENDAR AND THE MERIDIANS

The height of the sun at noon is the highest it reaches that day, and an accurate determination of date can be made by measuring this – the sun's culmination. It may be surprising that Cleopatra should be mentioned twice in a book on sundials. When Julius Caesar met her, she was twenty-four, beautiful, without scruple and fiercely ambitious. It is reported that she introduced herself to him by sending him a present of a fine carpet, into which she had had herself rolled up. He was fifty-four, away from his wife, and the most powerful man in the world. As might have been said of their meeting, the rest is history. However, during his passionate affair with her he also found time to develop the first calendar of reasonable accuracy. The Greek scientist Sosigones, whom Cleopatra had introduced to Caesar, had advised him to modify the calendar of the known world and introduce into every fourth year an extra day. So the leap year was invented; henceforth the Julian calendar would be followed, a reasonable approximation and a great improvement on previous arrangements. It was known to astronomers that the year was some minutes shorter than 365¼ days, as a result of Hipparchus' brilliant reasoning, but no one worried about this discrepancy. At the time there was no need for much greater accuracy, but later there would be, and that accuracy could only be achieved by even more accurate measurement of the sun.

The next landmark in the history of the calendar was the determination, by Christian bishops under the authority of the Emperor Constantine, of the date of Easter. Constantine was not baptized until on his deathbed, but in order to consolidate his power he had previously adopted Christianity as the empire's state religion. In 325 he summoned the leaders of Christendom to Nicaea, a small town to the south of modern-day Istanbul in Turkey. The location, also the birthplace of Hipparchus, is now known as Iznik and the ruined foundations of the building where the 300 or so bishops and their delegations met can be seen. Some bore horrific injuries from torture and persecution for their faith. Since its inception Christianity had lacked a central authority. Being scattered and often hounded by the authorities as subversive it had operated as a series of separate sects following the same basic tenets but inevitably after 300 years with no central command the various groups had started to differ on many points of practice and doctrine. Nicaea was the first council in the history of the Christian church, convened principally to address these differences, which had led to schism and were obscuring the message of the church, thus undermining the power of Church and Emperor. But the council passed other resolutions, which at the time seemed of lesser importance but to most of the world are of far greater relevance today. Previously there had been wide disagreement about the date on which to celebrate Easter. The difficulty for Constantine was to get everyone to celebrate Christ's resurrection on the *same* day, even if that day was not certainly known. It had to be fixed by his council as best they could. It was the most important date in Christian dogma and the timing of many other feasts was dependent on it, as is still the case. No one had recorded the date carefully. The Gospels were certainly not much help, as St John's account differs from the other three. Politically it was crucial to get agreement on this, for the state religion had to have one set of rules, and without a consensus Christianity could have foundered on uncertainty. It was known that Christ's resurrection had been during the feast of the Passover, and the Jewish lunar calendar had been in use in Jerusalem. The first month of the Jewish year was then normally the one beginning at the spring equinox, and called Nisan. Jews celebrated the feast of the Passover on 14 Nisan, when the Paschal lamb is slaughtered, and when, according to St John, Jesus was executed. The crucifixion took place

Pope Gregory XIII being addressed by the Commission for Calendar Reform (July 1582–June 1583), Archivio di Stato Sienna, Italy. The bearded figure pointing with his wand towards the sign of Scorpio may be a portrait of Christopher Clavius.

SCIPIO TVRAMINVS CRESCENTII FILVIS CV FVERIT MAGISTRATVS BICCHERNÆ
CAMERARIVS TEMPORE QVO GREGORIVS XIII PONTIFEX MAXIMVS ANNO REFORMAVER

on a Friday, his resurrection on the third day. On or about 14 Nisan the moon was full. At Nicaea several solutions were considered, one of which was to link Christ's resurrection to the solar year and to the Julian calendar by using the spring equinox as a fixed astronomical date. So it was decreed that the celebration of Easter would be on the first Sunday after the first full moon following the spring equinox, but scientists at that time could not predict precisely when the equinox would fall. Accordingly, most churches fixed on the date of 21 March, the date on which the equinox fell in 325. With some modifications, Easter is still defined in this way today. But another problem remained: the year was some 11 minutes shorter than the length decreed by Caesar, so the equinox would inevitably move one day earlier every 130 years or so.

The problem rumbled on. By the sixteenth century, after 1,200 years, the equinox was falling on about 12 March, but Easter was being celebrated as if the equinox was still to occur nine days later. Many devout scientists were shocked by this mockery of sacred dogma. Successive popes had been made aware of the problems over the centuries, but it was not until Pope Gregory XIII that action was taken. In 1582 he appointed a commission of learned scientists and worthy men to consider the problem and report, ten people in all. None of their names have become famous, but two of them are worthy of mention. Antonio Lilius, with his brother who had earlier died, must be given most of the credit for the solution to the problem, and Christopher Clavius, a Jesuit astronomer renowned in his time, managed to persuade the Pope to accept the proposed solution. This was to keep the leap year every four years but not for the century years that are not divisible by 400. This is the leap century rule, which drops three days from the calendar every 400 years by cancelling the leap year in three out of four century years. The leap year rules were a victory for common sense over scientific purity. Some members of the commission argued that a year was a year, however inconvenient its length might be, and would have been happy to have a calendar with a year of about 365.242 days (or whatever figure was considered most accurate at the time), subtracting a day by Papal Bull every 200 years or so to put matters right. But the success of Gregory's rule is that it has since endured without any modification. It can be stated in a relatively simple way and is a remarkable solution to a complex problem, because it leads to a calendar running ahead of the seasons by only one day in every 3,300 years. The other proposal, which the Pope also accepted, was to remove ten days from the calendar to bring the spring equinox back to the date it had been at Nicaea.

The Catholic world was ordered by Papal Bull to adopt the Gregorian calendar, which is now universal in the world for civil purposes. Others were not in such a hurry. Even today the Greek Church still defines Easter according to the Julian calendar, and so celebrates it on a different Sunday from all other Christians. Of course the Protestant world did not accept the ruling. Elizabeth I declined to follow. She almost certainly accepted the scientific arguments, but wished to avoid stirring any pro- or anti-Papal pot. It was not until 1752 that the British Parliament enacted the change, and by then eleven days had to be subtracted due to the fact that, under the Gregorian calendar, a further day was lost because 1700, by Gregory, had not been a leap year.

There was the usual uproar, which any great change induces in a country that enjoys democracy, but there was a particular reason for concern. What about the eleven days' rent which would be lost or, much worse for any government, eleven days' tax? It happened that in England before 1752 a government accounting period started on 25 March (Lady Day). In the reform, Parliament changed the start of the year to 1 January, but to overcome the financial problems of losing tax and rent, 11 days were added to the traditional accounting date of 25 March, which explains why the tax year in Britain starts at midnight on 6 April.

Readers may wonder what the calendar has to do with sundials. The first answer is that the Gregorian reforms were celebrated by the construction of a very beautiful sundial in the Vatican. Secondly, among the astronomers in the Catholic priesthood a renewed desire was developed to calculate the length of the year with even greater accuracy. With such information they could predict many years ahead with certainty when the equinox, and hence Easter, would fall. Giant sundials were needed to make these measurements.

Pope Gregory reigned from his election in 1572 at the age of seventy until his death thirteen years later. During his early ministry he had experienced in the raw immense upheavals in Christendom. In 1527 Martin Luther broke with the Church of Rome, and Christendom was suddenly divided into irreconcilable factions. The Protestant challenge to Rome's authority and supremacy rocked the Christian world. As we have seen, a calendar that did not harmonize with the most important date in the Christian past was seriously deficient, and undermined the Catholic Church's claim of sole ownership of the origins of Christian faith. Not only did Gregory reform the calendar during his short pontificate but he had the responsibility for mending the fractured institution of the Church itself. To this end one task he undertook was to restore Rome's churches and public buildings. A major monument he commissioned was an apartment of seven rooms in a three-storey tower in the Vatican. On the floor in one of the beautifully decorated rooms of this building is a meridian line, installed to celebrate the most famous achievement of his papacy, the calendar reform. This meridian is operated by a ray of light emitted from a hole which is in the mouth of a heavenly figure – presumably the Holy Father, reclining in the clouds on the frescoed south wall.

The meridian, an accurate sundial for measuring noon only, works like this: the projection of sunlight passes across the floor and at the instant of noon crosses a strip of bronze inlay, which has been accurately inserted precisely along a north–south line. Such an instrument also tells the date, read from dates engraved on the bronze, for the sun casts its longer winter beam further from the frescoed wall than its summer beam. Naturally the dates of the equinoxes are shown according to the Gregorian reforms of 1582, and this meridian established soon after is the first date-telling sundial that is calibrated in accordance with them. These instruments are similar to the noon dials we have already met, such as the one in the Radcliffe Observatory. They differ, however, in that they tell Apparent Solar Time, not Local Mean Time.

Egnatio Danti, the Dominican priest and mathematician who, under Clavius, was the second most important member of the commission for calendar reform, was the scientific inspiration behind the decoration, and the painter was Niccolò Circignani. The Pope wanted these rooms to celebrate not only his calendar but also the success of his pontificate. Danti was not only a priest and scientist, but also an amateur artist and architect and was supremely well qualified to interpret these wishes. The rooms were named the Tower of the Winds, a classical allusion to the famous building in Athens (see pages 23–5). In the ceiling vault at the centre of the meridian room he designed, inspired by Athens, an anemoscope (a wind direction indicator) surrounded by frescoes depicting personifications of the winds. It is inscribed with the Greek and Latin names of the winds and the indicator points to twelve representations of the winds seated on clouds below. Each of the four major winds is flanked by two lesser winds, and again by even lesser winds so that thirty-two personifications are shown grouped around the wind vane. Below, as framed pictures, are the four seasons, from spring in the form of a young girl to an old man for winter. In fact the many lovely frescoes in the rooms were designed to symbolize the domination of Christianity, in the form of Pope Gregory, over nature, time, and past and present cultures – as demanding a commission as anyone could make.

LEFT *The Calming of the Tempest* (c.1582), fresco by Niccolò Circignani, Tower of the Winds, Vatican. The spark of light from the opening in God's mouth can be seen in the upper centre right. This ray of light falls on the floor and crosses a bronze inlay at the instant of noon, sun time.

The fresco depicts the miracle of the Calming of the Tempest by Christ on the Lake of Galilee (Matthew 8:23–33). The terrified disciples can be seen huddled in the boat pleading with Christ. The same passage in Matthew's gospel describes the miracle at Gerasa

when Christ drove the devils out of two lunatics and into a herd of swine, illustrated on the lakeside. In this fresco we are made aware of Christianity's power over some of the major destructive forces facing mankind: time, violent storms and insanity.

But there is another message. The fresco is on the south wall and faces north. It also symbolizes the battering, faced by Christ and his disciples in Peter's boat, by the destructive and mad schisms emanating from the north, namely Martin Luther and Protestant heresy.

ABOVE The vault (c.1528), fresco by Niccolò Circignani, Tower of the Winds, Vatican. The anemoscope is operated by a wind vane on the roof which is connected by rods and gear wheels with the pointer in the centre of the ceiling.

To a sixteenth-century believer, measuring time was a way of knowing how many years would be left until the Judgment Day, until the end of time itself. Gregory's reform was regarded as a sacred activity to re-establish divine order in both the Church and Nature. To determine the calendar more accurately was therefore not just to fix the date of Easter or to improve man's knowledge of astronomy, but to advance the triumph of Christianity itself. For the next 200 years such meridian lines were built in churches and public buildings all over Europe. At least ten survive in public buildings in Italy and examples are to be found in other European countries. They existed in some of the larger private houses also, as a means by which the so often inaccurate household clocks could be reset.

Another particularly fine one is to be found in one of Rome's most famous churches. The baths of the Emperor Diocletian were completed 1,700 years ago, after an army of workmen had laboured for eight years to build a complex that covered more than thirty acres. It provided hot baths, swimming pools, gymnasia, libraries and areas of relaxation and entertainment for the daily enjoyment of thousands of Roman citizens. Sport and culture were combined to develop, as the poet Juvenal put it, '*mens sana in corpore sano*' (a healthy mind in a healthy body). The large central hall, with eight immense columns of red granite that were floated down the Nile and across the Mediterranean from Egypt, still stands. Its roof is intact, the interior vaulted with giant arches and lit from massive half-moon windows, termed by today's architectural historians 'Diocletian windows'. From outside the building is unimpressive, save for its size. Its façade appears as a mass of rough brickwork from which emerges a not infrequent weed, like many a romantic view of ancient Rome.

Inside, however, it is magnificent. In 1561, a Sicilian priest Fr. Antonio Lo Luca had a vision of a host of angels swarming round the building, confirming to him the tradition that the original builders had included 40,000 Christians in forced labour, many of whom died in the misery of their work. So in 1561 Pope Pius IV commissioned Michelangelo to build inside the structure a church dedicated to the angels and these early martyrs, the basilica of Santa Maria degli Angeli e dei Martiri. This masterpiece has been developed and enhanced ever since by architects, artists and one brilliant scientist, humanist and priest. Francesco Bianchini was born in Verona in 1662. He was commissioned by Pope Clement XI to create another giant meridian line in the city. If measured and constructed with great care such a sundial would tell the time of noon with accuracy and provide a fine instrument for other astronomical observations.

Why did the Pope order and pay for this, and in the year 1701? Since the reform of Gregory XIII in 1582 the Papacy had retained the power to regulate the calendar for the Catholic Church. Furthermore, under the reform the year 1600, being divisible by 400, had been a leap year, but the year 1700, not being divisible by 400, was not a leap year and it was therefore the first centennial year since Julius Caesar that had not been. To the ignorant public, and to the educated but unscientific, this was a matter of concern because it could lead to the celebration of Easter several days later than they considered right. The problem of when to celebrate Easter was still troubling the faithful. Again the Universal Church had to try to bring certainty to this area of doubt. The dates of Easter and of all the other movable feast days that derive from it were dependent, as we know, on the timing of the spring equinox. It was understood how to anticipate this reasonably well but the new moon always occurred on different days in different parts of the world. So it was important to ascertain an even more accurate understanding of the moment of equinox, and be able to predict years in advance to within minutes when it would happen, for it might happen a few minutes just before midnight in one place or just after midnight in another.

Cannon dials, French, nineteenth century. In Rome, until 1846, at noon read from the dial in Santa Maria degli Angeli, a cannon was fired to enable the citizens of Rome to adjust their clocks. A novel form of dial is the cannon dial. The instrument was charged with gunpowder and ignited by the magnifying glass at midday. This example follows a tradition of such dials. It was never very practical and was more of a scientific curiosity, or grown-up's toy.

Unless one knew which day for certain, there might be problems like the celebration of Easter on a different day in different parts of the Catholic world, which now stretched over the entirety of the globe far beyond Europe. An instrument that could demonstrate an even more accurate explanation for the recurrence each year of this event would give confidence to the faithful, and a great meridian could measure the time of the equinox as well as noon. Of course the Pope's intentions may also have been coloured by vanity. All powerful leaders wish to leave behind memorials to their achievements, and Pope Clement may have thought that there was no better centennial monument than a giant sundial. Architectural projects always please the powerful. Probably also his advisers, many of whom were Jesuit priests with a scholarly tradition of studying astronomy, pressed upon him this project for their love of science.

Bianchini decided that the great basilica was the ideal site. The building was very solid indeed. It had already stood for 1,400 years and its foundations and walls had not been moved even by earthquakes; the instrument would be stable. And the church was vast and dark. So the Gnomone Clementino, as it was known, was constructed. A bronze strip, a couple of inches wide and nearly 130 feet long was inserted into the then brick floor along a north–south line. High above it, at 66 feet, is a hole set into the wall and surrounded by a giant carving of the Pontifical arms. At the instant of noon sun time an image of the sun crosses the centre line of the strip. At the midsummer solstice the sun image appears almost circular. By the time of the autumn equinox it is clearly elliptical and by midwinter it forms an ellipse measuring 43 × 18 inches, projected from the hole which is only 3/4 inch in diameter. What you see is not an image of the hole but of the sun itself, which is thrown in a cone of light for a distance of 164 feet on to the floor.

The thirteenth-century English savant Fr. Roger Bacon had anticipated the need for Pope Gregory's reform by three centuries. He is also credited with inventing both spectacles and the camera obscura. Clement's meridian is, in fact, another giant camera obscura. At the time of the eclipse by the moon of the sun on 11 August 1999 the shape of the moon obscuring part of the sun image was clearly visible. Surrounding the meridian are some exquisite marble inlays, intarsi representing the twelve signs of the zodiac, the work of a master. These were carved, according to the custom of the time, from pieces of colourful marble, such as giallo di Numidia and rosso antico, recently excavated from archaeological sites but which had been extracted in antiquity from different places in the Roman empire. On each of the beautiful carvings, shown in detail on page 31, are inlaid bronze stars conforming to the layout and magnitude of the stars and nebulae of each zodiac.

Until 1846 this device was used to regulate the clocks of Rome but because of its immense dimensions and careful construction it had also been used to measure accurate times for the instants of the equinoxes and solstices. At places on the meridian near where the sun must cross at the equinoxes are bronze scales with very detailed engravings. Scientists of the period could measure the moment of the equinox with accuracy, certainly to within less than a minute. On the last noon before an equinox Bianchini measured from these scales the distance along the meridian of the sun's transit with the greatest care, probably to within a fraction of an inch on the floor. Next day, at the first noon after the equinox he made the same observation. The equinox would have occurred between these two events, and with a complicated procedure it was possible to calculate when. By doing such observations over several years and averaging the results great accuracy was obtained. This enabled him to calculate the length of the year at 365 days, 5 hours, 49 minutes and 1.5 seconds, sixteen seconds out according to modern calculations.

SANTA MARIA DEGLI ANGELI E DEI MARTIRI

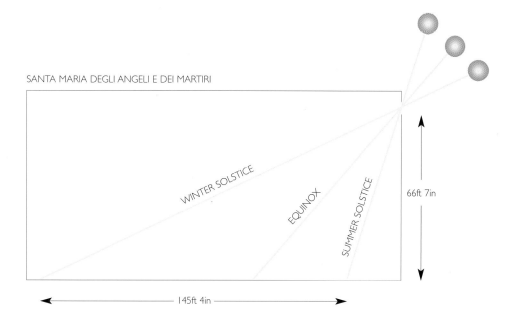

WINTER SOLSTICE

EQUINOX

SUMMER SOLSTICE

66ft 7in

145ft 4in

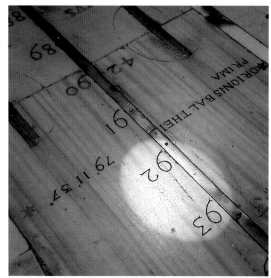

ABOVE This diagram explains the different positions of the sun's projection. At the winter solstice the sun image is projected 164 feet from the wall.

RIGHT A view of the interior of Santa Maria degli Angeli, from *De Nummo et Gnomone Clementino* by F. Bianchini (Rome, 1753). Bianchini wrote this book to explain the making of his great meridian. The sundial is the shaft of sunlight on the right projecting from a hole in the papal arms down to the floor and is marking noon at the summer solstice. The shaft on the left is imaginary and represents a line through an opening, in the centre of a cross, which leads to the Pole Star. With the aid of a telescope placed in the meridian on the floor the star was observed and measurements were made to determine a more accurate value for the Precession of the Equinoxes.

LEFT CENTRE The sun image crossing the bronze strip a few second before noon.

LEFT BELOW The gnomonic hole in the arms of Pope Clement.

CLEMENTIS XI Pont.Max; ad usum Paschalis Festi et Kalendarii · Gnomon Meridianus constituitur Iº ex radio verticali A D extenso palm·Rom·91·unc·²/₄ siue pedis Regii Paris·unc·750·qualium ra
cto perpendiculi D ad extremum F quo pertingit solaris radius in Tropico Capricorni· 7º ex hypothenusa A V C, quam radius solaris efformat· Gnomon Polaris, altus uncias pedis Regii 900· ex umbilico B ad pavime
lis circuli H I cum pavimento ostendit latitudinem Vrbis hoc in loco grad·41·5·4· 27· F extremum Tropici Capricorni· C extremum Tropici Cancri: unde colligitur ex observationibus collatis declinatio Tropicorum grad· 2
in paralleli diurni circa polum· N O Parallelus diurnus Arcturi· Q R Parallelus diurnus Sirii: que stelle per tubum opticum etiam perdiu videntur ex umbilico A, ubi fenestra aperitur, immoto Gnomon
di in Thermis, et hanc proximam S.MARIAE Angelorum cum inviseret, primum inspexit· Idem parallelus incidit in declinationem termini Paschalis Y diei 25·Aprilis, ibi notatam· Observata sunt per Gnom
uv·h·8·m·45·p·m·n· Æquin·Autumn·die Dominica 23·Sept·h·9·m·4·p·merid· Solst·Hybern·Sabb·die 22·Decemb·h·XI·m·45·post·med·noctem· K Γ una ex 100 partibus totius altitudinis A D

FURTHER EAST

Islam had resupplied Europe with the science of sundials and also spread the learning eastwards to Asia. It has been written that when the Maharajah Sawai Jai Singh II of Jaipur died in 1743 his wives, concubines and science expired with him on his funeral pyre. There is no doubt that this unusual ruler was inspired at least in part by Ulugh Beg in his love of giant astronomical instruments. He possessed in his library a copy of Ulugh Beg's star tables, which he had his assistants bring up to date.

He owed allegiance to the Emperor Muhammad Shah, and dedicated his compilation of astronomical tables, the *Tij Muhammad Shahi*, to him, because the tables were compiled with his permission and ostensibly for the better administration of the empire. He wrote in his dedication:

> To his Majesty of dignity and power, the sun of the firmament of felicity and dominion, the splendour of the forehead of imperial magnificence, the unrivalled pearl of the sea of sovereignty, the incomparably brightest star of the heaven of empire, whose standard is the sun, whose retinue the moon, whose lance is Mars and his pen like Mercury with attendants like Venus, whose threshold is the sky, whose Signet is Jupiter, whose sentinel Saturn – the Emperor descended from a long race of kings, an ALEXANDER in dignity, the shadow of God, the victorious king MUHAMMAD SHAH: May he ever be triumphant in battle.

In likening the emperor to all the known planets as well as Alexander the Great, he disclosed a remarkable gift for flattery and metaphor, but there were possibly other reasons for this exuberant language; his position was never wholly secure and he survived only by means of great political cunning.

Jai Singh had succeeded to the Amber territory at the age of thirteen and throughout his life was a stargazer with a passion for astronomy. A long period of disorder coincided with his reign but he was a successful statesman during this very dark period of Indian history. He was a

BELOW *The Samrat Yantra, Delhi,* lithograph from *Oriental Scenery* by Thomas and William Daniell (London, 1808). Thomas Daniell was an English landscape painter who went to India in 1784. He took with him his fifteen-year-old nephew William as an assistant. The two of them spent ten years painting views all over the continent and on their return they worked upon their important publication, *Oriental Scenery*, which contained 144 engravings. The book was a commercial success for eighteenth-century Britain was fascinated by these views of its exotic and greatest colony.

prodigious builder who founded Jaipur and built giant observatories in five cities – Delhi, Jaipur, Benares, Ujjain and Mathura. His instruments were of marble stone and lime, fixed, sometimes gigantic, and of a simple geometric form, which later was to inspire Le Corbusier in his designs for the Indian city of Chandigarh. There was a whole range of instruments at each location: giant sundials, bowls like the hemispherium of Berosus, other giant bowls in the ground for star measurement, pillared circular buildings for measuring the direction of stars, the sun and planets (known as their azimuth), and many other structures. But the pride of them all are the Samrat Yantras (supreme instruments) at Delhi and Jaipur.

These beautiful objects, which seem so modern, resembling contemporary architecture, were useless for the accurate measurement of the heavens and bore no comparison to European instruments with telescopes and the accurate scales that were available. Jai Singh could have purchased these items because he was in contact with several European scientists, including a Portuguese Jesuit priest, who assisted him. Following Islamic precedents, the Maharajah had believed that large instruments were more accurate than smaller ones, and that brass instruments with moving axles which wore down were inferior. However, by making fixed structures of stone he introduced inflexibility into them and failed to appreciate that large sun instruments produced shadows that were fuzzier than small ones and thus gained little. His astronomy did nothing original, and was confined to confirming measurements that had been made by earlier Islamic and Greek savants.

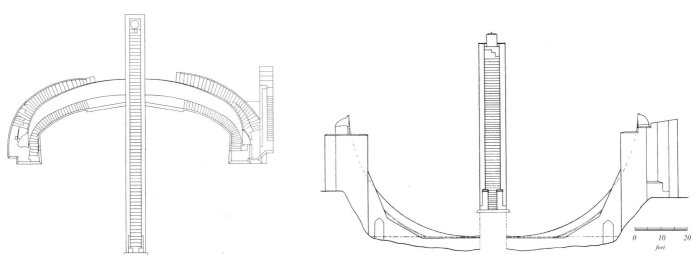

LEFT ABOVE A view of the Samrat Yantra at Jaipur, a huge equatorial sundial whose gnomon bears a staircase with more than fifty steep treads.

LEFT BELOW Plan and elevation of the Samrat Yantra at Delhi, from *The Astronomical Observatories of Jai Singh*, by G.R. Kaye (New Delhi, 1918).

RIGHT ABOVE Hemispherium, Jaipur. Among the many instruments at Jaipur is this limestone hemispherium. It measures about 8 feet in diameter and readings can be taken from a spot of light projected from a ring which is suspended from cross wires.

RIGHT BELOW Homage to Jai-Shing II, 1995, by Ono Yukio, height 6½ feet. Ono Yukio demonstrates his inspiration from the past in this sculpture in which art and science are clearly combined. Twelve divisions of the day are projected on to a curved surface in a way chosen by the sculptor to produce engraved curves of harmony.

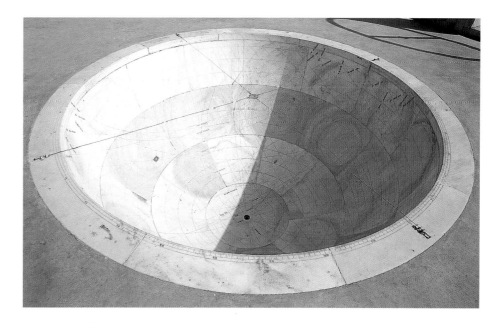

There is one scholarly theory which claims that these instruments were concerned less with accurate measurement and more with Indian mysticism, and that Hindu priests and mystics, as well as mandala theory, were the main influences on these projects. The monuments were visible from afar outside the cities and were therefore a visual demonstration of the ruler's great earthly power and a display of his cosmic knowledge and superiority over his illiterate people.

As we have seen, Jesuits had been at work in India. There was a thriving group from the Society of Jesus in Goa. In the seventeenth and eighteenth centuries Jesuit scientists both travelled and investigated the world about them incessantly. Often polymaths, they excelled in a rich variety of scientific fields: physics, mathematics, geography and natural history; but their greatest achievements were always in the field of astronomy. In 1651, Giambattista Riccioli SJ published his *Almagestum Novum* (*New Almagest*), harking back to Ptolemy, and it contained a map of the moon, which became a standard of the time. He thus succeeded in immortalizing 35 Jesuit astronomers by naming lunar features after them. These included, of course, Christopher Clavius, whose lunar crater would be made famous as the site of the appearance of the mysterious monolith in Stanley Kubrick's film *2001: A Space Odyssey*.

Islamic science went even further east and via central Asia reached northern China. Here, it was later augmented by the knowledge brought by missionary priests from Europe. The Beijing observatory was built in 1442 and on its roof is a range of astronomical instruments including a sundial supported on exotic bronze dragons. These date from the eighteenth century and were introduced by the Jesuits. From the 1580s, Jesuit scholars travelled as missionaries to China, where they spread their scientific knowledge as well as religion. They introduced the newly invented telescope to the Chinese, but protected them from the controversial Copernican theory of the sun, and impressed on them the earth-centred system, a view that prevailed in China even into the nineteenth century. Jesuits had been allowed by successive Emperors to work in China in return for the supply of astronomical information, like the dates of eclipses in advance. Such information was used by the Imperial Court to emphasize the Emperor's divine status and especial knowledge.

The first British Embassy to the Celestial Kingdom, that of Lord Macartney, took place over several years which included 1789, the date of the French Revolution. Although the Society of Jesus had originally sent astronomer priests to Beijing, by the late eighteenth century they had begun to send people with no such expertise, for the information required by the Emperor was available each year in a French ephemeris, *Connaissance des Temps*. However, the turmoil in revolutionary Paris had led to the cessation of this publication. The poor priests, knowing they could not supply the information the Emperor desired and in fear for their lives, begged the scientists in the Ambassador's entourage for the information. English decency and Christian charity prevailed, and the scientists duly supplied the information to priests who were both Catholic and from a fiercely rival country, which the English would shortly be fighting for their survival.

There would seem to be a theme that linked several cultures and ages. Christianity needed sundials to anticipate Easter, Islam to regulate the times of prayer. Hindu and Chinese potentates, as well as Augustus, used them for mystical and religious reasons and to impress their heaven-given power on their subjects. Even Stonehenge can be linked with Santa Maria degli Angeli e dei Martiri and the Vatican's Tower of the Winds. They were all the creation of astronomer priests. The mystery of the cosmos had its connections with both mysticism and religion. Such ideas, however, did not survive in Europe beyond the sixteenth century, for soon scientists were to hold that all knowledge and progress were the product of informed and enlightened reasoning.

RIGHT The Astronomical Observatory in Beijing dates from 1442, in the Ming Dynasty. Until the twentieth century it stood in open countryside.

BELOW LEFT The New Armilla, 1744, Qing Dynasty, Astronomical Observatory, Beijing, China. The instruments on the roof of the observatory were designed by Jesuit priests. Though remarkable and fine examples of casting, they were inferior to the contemporary instruments of Europe. The New Armilla was used for solar measurements as well as other astronomical observations.

BELOW RIGHT The Royal Hong Kong Golf Club equatorial dial, 1989, by Joanna Migdal, phosphor bronze, 4 feet diameter. This dial was commissioned as a centenary present from the Royal Hong Kong Jockey Club. The design brings together British and Chinese traditions. It bears Western and Chinese numerals, years, zodiacs and other elements, including race horses, and is mounted on two bronze dragons. After its initial installation, the Chinese feng shui expert insisted that its position was inauspicious, and it was relocated.

9

THE AGE OF ENLIGHTENMENT

Portrait of William Oughtred. His face displays intelligence and modesty. It is no surprise that he was much loved by his contemporaries.

One of the most beautiful types of portable sundial must be the universal ring dial. Although it is clearly derived from the astronomical ring of Gemma Frisius, which has been mentioned, it was developed by an Englishman, William Oughtred. Oughtred was a highly regarded mathematician who was active in the early 1600s. The son of the registrar at Eton College he was born and educated there. He was friends with many of his contemporary scientists and, both from accounts of the time and from his portrait, appears to have been a most kindly and learned man. Sir Isaac Newton spoke of him as 'that very good and judicious man, whose judgment may be safely relied upon'. His invention is closely related to the armillary dial, the astronomical ring and the equatorial dial. Its advantages are its portability and that it can be set up without a magnetic compass. It can be adjusted for any latitude by means of the ring at the top. In accordance with the basic rules, the dial will be accurate provided the time ring is parallel with the equator and the gnomon, which is at a right angle to the time ring, is parallel with the earth's axis. In this case the gnomon is not a rod but a flat plate with a slot, in which there is a sliding cursor pierced with a hole. Through this, the sun projects a spot of light. The cursor is adjusted up or down to the correct date, which was usually marked according to zodiac and calendar months. The dial is suspended from the ring at the top and rotated until the sun image falls on the time ring. The height of the sun at any hour varies according to season, and the date markings are calculated to align the dial correctly in the north–south position when the sun is projected as described.

There was considerable advantage in avoiding the use of a magnetic compass to set the dial because compass readings need adjustment for they suffer from magnetic variation, which differs all over the world and also changes over time. It was discovered in the 1630s that the error could change in one place as much as 5° over fifty years. For a traveller, therefore, the Oughtred dial had many advantages.

Oughtred had caught the attention of the Earl of Arundel, and became tutor to his son. In the 1630s the Earl and his wife planned an expedition to Madagascar. It was never realized but in

BELOW Universal equatorial ring dial, from *The Book of Sundials* by Mrs Alfred Gatty (London, 1900) and modified. In this illustration it can be seen that the latitude is set at the top for about 52°. It reads 3pm. This image is taken from a well-known old book on sundials. It shows how the hour ring can be folded flat into the latitude ring so that the instrument can be carried safely in its box. Note the latitude information, from Rome in the south to Amsterdam.

RIGHT Universal equatorial ring dial, by Thomas Heath, London, c.1740, 12 inches diameter. A fine example of the many dials that were made to this design. The outer scale is engraved with the latitudes of numerous European cities and it is suspended on a fine gimbal-mounted loop.

BELOW RIGHT Universal equatorial ring dial, by Thomas Heath, London, eighteenth century, 9½ inches diameter. The ring for adjusting the latitude is clearly visible at the top, set for latitude 46.5° north, as is the pierced cursor, set for 12 March/ 11 September. Craftsmen continued to make these ring dials into the nineteenth century.

LEFT *The Earl and Countess of Arundel 'The Madagascar Portrait'* (1639), Anthony Van Dyck, Arundel Castle, Sussex. The Earl, known as the 'collector earl' because of his passion for art, had other ambitions as well. He became a Governor of the New England Company, and when Van Dyck painted this portrait he and his wife were involved in a scheme to colonize Madagascar. The venture collapsed because they heard the island was infested with fleas.

FAR LEFT Crescent dial, Augsburg, 1720, by Johann Wildebrand, silver gilt and steel, c.4 × 4 inches. This very fine example of craftsmanship is an unusual dial which is derived from the universal equatorial dial. Instead of a ring, there are two semi-circular arcs marked with numerals and a crescent-shaped pointer for each dial. In use, these pointers are folded upwards so that the adjustable date scale is perpendicular to the plane of the two arcs. The dial is adjustable for latitude and fitted with a plumb weight and two screws for levelling. Johann Wildebrand was the last of the great Augsburg makers.

LEFT The sundial designed by Sir Christopher Wren at All Souls College, Oxford.

anticipation of it Van Dyck was commissioned to paint *'The Madagascar Portrait'*. The Earl is shown pointing to the island on a globe and the Countess with a pair of dividers in one hand and an Oughtred dial on her lap, which surely indicates the high esteem that he enjoyed in the eyes of his patron.

A younger contemporary of William Oughtred was Christopher Wren. Wren came from the same sort of professional background as Oughtred, though slightly grander. When he was three years old, his father was appointed Dean of Windsor and register of the Order of the Garter, jobs that came with a fine house and enabled the incumbent to rub shoulders with the highest nobility. Both father and son were ardent Royalists all their lives, but the family's prosperity and status were shattered by the Civil War of 1642. Young Wren was aged twelve. The trauma may have been a spur to his ambition, and certainly did not diminish his remarkably successful life, not only as an architect of world fame, but also as a significant innovator in science, medicine, mathematics and astronomy. Both Oughtred and Wren enjoyed a passion for sundials. As a teenager Wren had constructed sundials all over his father's house. As an undergraduate at Wadham College, Oxford, he depicted a sundial on the ceiling of his room illuminated by a spot of light reflected from a mirror on the windowsill. Maybe one day someone will scrape the paint off the ceiling of a south-facing room at Wadham and discover it. He was connected with William Oughtred. While only fifteen he translated into Latin a treatise on sundials which the learned mathematician had written. This was published in 1652 with a preface by Oughtred in which he referred to the translator's skills:

> A youth generally admired for his talents, who, when not yet 16 years old, enriched astronomy, gnomics, statics, and mechanics with brilliant inventions, and from that time has continued to enrich them, and in truth is one from whom I can, not vainly, look for great things.

This work received considerable critical acclaim and was one reason why he was granted a fellowship of All Souls College at the unusually young age of twenty. Wren immensely enjoyed his fellowship at All Souls. In 1659 he designed the great sundial that is to be seen over the centre of the south front of the Codrington Library. It was moved to this position in 1871 and it is to be hoped that it will one day be returned to its original position in the centre bay of the south wall of the chapel.

Another man of genius who developed a passion for sundials at a very young age was Isaac Newton. When we say that our Government has lost momentum, or that we resent the inertia of bureaucracy, feel the force of gravity, or talk about motion, of mass, of action and reaction we are unconsciously using words that were given enriched meaning by Newton, the first man of science to be honoured with a knighthood. It is always moving when a strange, lonely man, the son of an illiterate farmer, suffers a poor and unhappy childhood, yet struggles and manages to develop his full potential and gain the enduring recognition by his generation and all subsequent ones of genius.

He studied secretly, obsessively and without break, and had penned more than a million words before publishing anything of significance. He wrote for himself, and the majority of his lifetime's writing was on alchemy, not science. Yet he was a man who sought and believed in order. He was respected but feared and resented by nineteenth-century poets and artists for explaining the secrets of the prism and hence spoiling their romantic view of the rainbow. He had tried to 'conquer all mysteries by rule and line' according to Keats' poem 'Lamia' (1819). Wordsworth had seen at Trinity College a statue by moonlight, which he described coldly:

I	2	3	4	5	6	7	8	9	10	11	12	13	14	15
0·48	1·36	2·24	3·12	4·0	4·48	5·36	6·24	7·12	8·0	8·48	9·36	10·24	11·12	12·0
16	17	18	19	20	21	22	23	24	25	26	27	28	29	30

NEWTON: AGED 9 YEARS: CUT WITH HIS PENKNIFE THIS DIAL. THE STONE WAS GIVEN BY C·TURNOR ESQRE AND PLACED HERE AT THE COST OF THE RT HON. SIR WILLIAM ERLE A COLLATERAL DESCENDENT OF NEWTON. 1877.

LEFT Not surprisingly there are many fine sundials at the famous seats of learning in Britain. Both Cambridge and Oxford boast them. The dial at Queens' College, Cambridge, is one of the most beautiful and elaborate. Its design has often been attributed to Newton, but this is erroneous for the dial was most likely built in 1733 and he died in 1727. The time is read from the shadow cast by the rod gnomon, the hours and quarter-hours being marked in black, but the gnomon holds a small gilded ball and it is from the position of its shadow that other readings are taken. The twelve signs of the zodiac are depicted, together with the months in Latin. These are bounded by curved bands painted green, so the zodiac and approximate date can be read. Red curves give the sun's altitude and vertical lines in black give direction, together with the established nautical bearings – for example, SWBS means south-west-by-south. The two vertical columns marked ORTUS SOLIS and LONGITUDO contain figures that indicate the time of sunrise and the length of the day in question.

Beneath is a table for telling the time by moonlight. You need to know the moon's age. The new moon is age 1 day, the first quarter is age 7–8 days, the full moon is age 15 days, and the third quarter age 22–23 days. On this dial you judge the moon's age by looking at it, then add the appropriate figure in the central boxes marked in hours and minutes to the reading. Then you must do four mental calculations. First you read the time which by moonlight will be faint, then judge the age of the moon and apply its correction, then apply your displacement from Greenwich, then the Equation of Time. No doubt the dons enjoyed explaining this game of the mind.

LEFT BELOW This dial was re-erected inside Colsterworth church under a bust of the great man. It was removed from a wall of nearby Woolsthorpe Manor in 1877 and placed upside down. Like the one at the Royal Society, this dial is crude and inaccurate and almost certainly was made by Newton in his early teens when he was untutored in geometry or astronomy, for, apart from a little arithmetic, Newton learnt no mathematics until in his early twenties at Cambridge. In many ways it resembles the early seasonal dials of Bewcastle and Chartres. There are many similar examples surviving on nearby churches. A recent study has conjectured that he made it by driving a peg into the wall and then, with a string attached to the peg, marking the hours by listening to the chiming of the church clock, and engraving the lines so formed to the peg, which becomes the horizontal gnomon.

ABOVE RIGHT William Blake must have hated science. In this famous image he depicts Newton sitting on a rock, concentrating on a geometrical figure of a triangle with an arc and measuring the base of the triangle with a pair of compasses, his mind obsessed by abstract reasoning and lost to the wonders of the natural world, the stars and the heavenly firmament.

Newton with his prism and silent face,
The marble index of a mind for ever
Voyaging through strange seas of thought, alone.

In his lifetime, Newton travelled no more then 150 miles from his birthplace, but at his death was one of the most famous people in the world. Alexander Pope had written: 'God said, "Let Newton be!" and all was light,' a joyful tribute, no doubt tinged with mischief, to a man who unravelled so many secrets affecting the sun.

It is interesting to reflect that even great men can indulge in pseudo-science. Newton did so with alchemy, and Kepler with Platonic solids. Sometimes an idea, a prejudice, is more important than the facts, which anyway are often uncertain, due to a lack of contemporary knowledge. Witness today the passionate scientific disputes about the environment, and also what we eat.

As a youth one of his passions was sundials. He was good with his hands and also made waterwheels and water clocks. He hammered wooden pegs into walls and the ground to measure time to the nearest quarter-hour and he cut sundials in stone. Neighbours near his home at Woolsthorpe Manor in Lincolnshire regularly consulted his dials. Two that he designed have survived; one is in the Royal Society in London and the other is now mounted inside Colsterworth church, close to his birthplace. The dial at Queens' College, Cambridge, often called the Newton Dial, is fine enough to be by Newton, but is not. His love of sundials remained with him all his life. It was said that he could tell the time of day from the position of shadows on the floor of his study.

It is not often that someone invents a new sundial, and when an invention has been made it has usually been in the form of a development from some earlier model, as was the case with Oughtred. Sometimes a development has taken place that shows a great degree of originality in its design. Over many years mathematicians must have wondered if it was possible to make an accurate sundial that had a vertical gnomon, the simple stick in the ground with which early

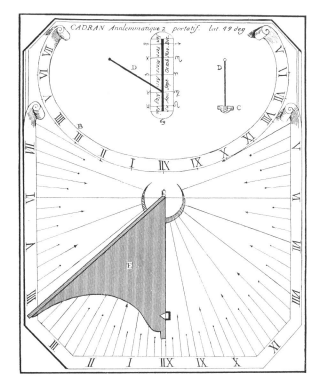

man started his crude measurements of the sun. At some stage, probably in France, the problem was solved and the first known text discussing the invention was written in 1640 by a now obscure mathematician called de Vaulezard. Very little is known about this man, which is remarkable since he lived in France during one of the greatest periods of innovation the scientific world has ever seen. He was an accomplished mathematician living in Paris, who wrote several texts on the analemmatic dial between 1640 and 1644, and also on perspective. A picture of him survives, but this is the sum total of our knowledge of this brilliant man. The type of dial he invented, with a vertical gnomon, is known as an analemmatic sundial, although the word analemmatic is problematic. Vitruvius first used it in a way that is obscure, but it is also used to describe the figure of eight curve in the Radcliffe dial. How the word analemmatic came to describe this dial is unclear.

There are two essential features that distinguish it from conventional horizontal ones. First, the hours and other times are not lines but points on a curve which is an ellipse. Secondly, the vertical gnomon must be movable to different positions on different dates according to a scale on the smaller axis of the ellipse. It is simplicity itself to look at but rather more complicated in its mathematical design.

LEFT BELOW Longwood Gardens, Kennett Square, Pennsylvania. In 1906 Pierre Samuel du Pont created one of the great and large gardens of the world at Longwood. In it he commissioned a giant analemmatic sundial, no doubt inspired by the original one at Brou in France but to which he had added a novel feature. As the gnomon must be moved to different positions according to date, why not incorporate in that movement an adjustment for the Equation of Time? The gnomon therefore does not move simply up and down as on a straightforward analemmatic dial, but sideways as well. It is impossible to do this with precision, for the theoretical position of the gnomon on any given date is not represented by a single point for all times of the day, but it is possible to make a close approximation and the average error from clock time of the Longwood dial for all dates is less than one minute.

RIGHT Readers are unlikely to want to make an analemmatic sundial, and there is therefore no point in being mathematical. The principles can, however, be envisaged by comparing it with the universal ring dial. The dial is an orthogonal projection of this dial, which means a vertical projection downwards with all the projecting lines parallel, as shown in red. The illustration is at two scales with the analemmatic dial enlarged below for clarity. Look at the hour circle on the ring dial. When projected it becomes an ellipse, while the pinhole and cursor of the ring dial becomes a vertical rod, which slides along one axis of the ellipse.

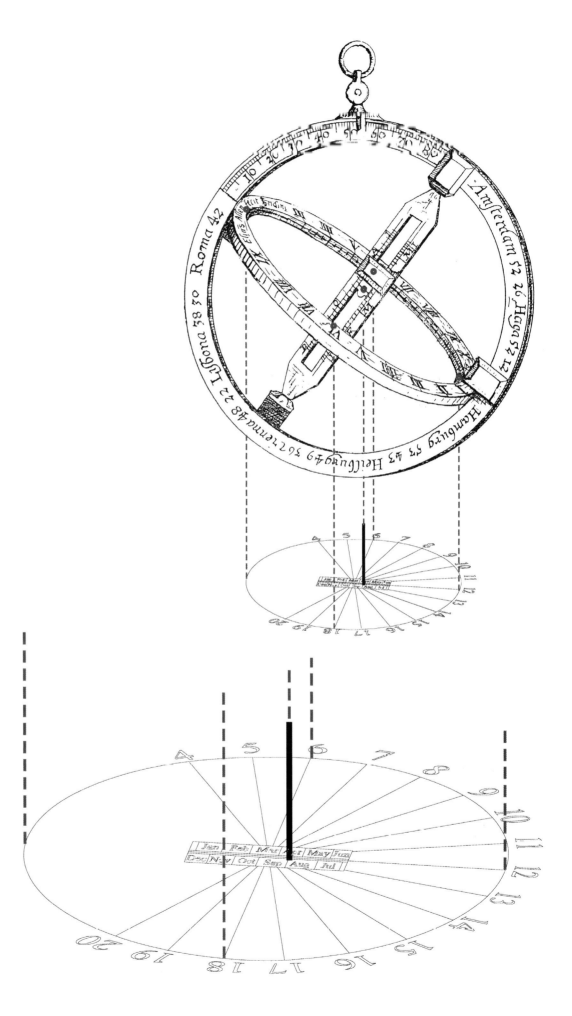

ABOVE English brass analemmatic dial, London, c.1740, by Thomas Wright, c.7½ × 6 inches. This instrument would have been used for setting household clocks to an accuracy of within about two minutes. It is a combined horizontal and analemmatic dial. The vertical edge of the horizontal dial gnomon is the analemmatic dial gnomon and the square plate on which it sits can slide to adjust the dial for date shown by calendar months and the zodiacs. On the inner rings of this plate are engraved values for the Equation of Time. It has adjusting screws and spirit levels to make the base precisely horizontal. It has one other important feature. The dial can be adjusted for different latitudes by tilting the plates against the arc-shaped arm at the end which has a scale of degrees (not visible) engraved on its far side. This makes the dial 'universal' and illustrates an important principle. Any plane dial calculated for a particular latitude will work in any other latitude provided the plane is tilted so that the slanting edge of the gnomon is parallel with the earth's axis.

RIGHT Sun compass, 1940s, Cole Pattern. A sundial must be pointed north to tell the right time. It follows that if you know the time you can use a sundial to find north. This is the principle of the sun compass. A magnetic compass is highly unreliable on any vehicle with rotating and electrical machinery. So during the Second World War in North Africa sun compasses were used regularly on military vehicles for navigation. In long range desert raids small groups of soldiers had to move swiftly and accurately across large distances with no landmarks for guidance. At night they used the stars, by day the sun. The SAS (Special Air Service), successors to the Long Range Desert Group, used sun compasses for desert navigation behind the lines in the first Iraq war. This instrument is based on the analemmatic sundial and could be used in latitudes up to 36° north or south of the equator.

Man often mimics the marvels of nature. When monarch butterflies undertake their annual migration from North American forests to winter in Mexico they are guided by a tiny internal compass and, even as the sun moves daily across their path, are always able to head south. Scientists claim to have discovered the connection between a light-detecting sensor in the butterfly's eye and nature's clock in its brain.

The earliest versions of this dial were large monumental structures, and the ancestor of them all is to be found beside the church of Brou, a village in eastern France near Lyon. At its widest point this dial is over 36 feet in diameter and the movable gnomon is a rod kept vertical by a tripod rather similar to the great dial commissioned in 1906 by Pierre du Pont in the garden of Longwood in Pennsylvania. Similar dials are also made today in public places, largely as an attraction for children, who can stand on the appropriate date and use themselves as the gnomon.

Soon after their invention these dials were transformed into instruments of great accuracy. Although it was obvious that they had one disadvantage – namely the need to know and set them for the date – they had one great and overriding advantage as well. They could be combined with a simple horizontal dial so that there were two dials in one instrument, working independently of each other. The whole assembly could then be rotated until both dials read the same time. The result was that both dials were then accurately aligned towards north, without the need for a magnetic compass, which never gives north accurately without complicated adjustments.

We are so accustomed to imagining our own age as one of great change that we fail to grasp the fact that our ancestors often had the same feelings about their own times. The seventeenth and eighteenth centuries in Europe experienced massive advances in invention and scientific development, as well as exploration of the world. Until shortly before the time of Newton and Wren many of the learned in Europe still believed, as they always had, that the Greeks had

discovered everything of importance in science, that there could be no further significant progress in astronomy, medicine or natural philosophy as science was then known. These people believed that the sum total of knowledge was finite, and that the well was full, or nearly full, to the brim. In the 1690s a fierce battle was fought between those who thought the knowledge of the ancients was still in general superior to that of the moderns and those who believed exactly the opposite. One of the arguments put forward by the moderns was the development of wonderful new technologies such as printing and the startling discoveries of Kepler and Copernicus, unknown to the ancients. This revolution in thought was stimulated in England by the challenge to the Crown and traditional thinking that resulted from the Civil War of the 1660s.

Newton and Wren thus lived and worked at the start of a period known to historians as the Age of Enlightenment, and throughout the eighteenth century an enormous interest in science was generated among the educated in England. It was an age when artists respected scientists and often portrayed them at work, of which there can be no finer example than the famous painting by Joseph Wright of Derby. London witnessed an explosion in the skills and production of precise scientific instruments. The demand was fuelled by new scientific inventions, by a growing interest in science among the educated and by the need for instruments of navigation for an empire growing larger and richer on its extensive maritime trade. An early example of one of these fine instruments is the dial constructed for Staunton Harold in Leicestershire.

Science was seen as a symbol of progress, of how to improve the condition of man through the use of reason. Wealthy patrons commissioned beautiful scientific models and instruments. When George III became king in 1760 he led the way as the grandest patron of all. He, as his father had done, commissioned from the leading instrument makers of the day the most magnificent instruments of all kinds including astronomical models, armillary spheres and sundials, all exquisitely engraved and made of the finest materials, the pride today of the Science Museum's collections in London.

For many years, the Royal Society, the learned institution that had been founded in 1660 under royal patronage for the pursuit of science, held lectures explaining contemporary scientific developments. These lectures were open to the public at large, and had attached to them a not inconsiderable entrance fee, such was the interest in science held by educated people. Some lecturers and popular writers found they could enjoy a modest income from this work. One person who did so has already been mentioned, James Ferguson, a fine example of a brilliant man from a very poor background, even more modest than that of Newton, who became famous in his lifetime. At the age of ten he went to work as a farm labourer, tending sheep by day and studying the stars by night. Though never well educated, he had the talents of a mechanical genius, and could describe such things clearly and simply to people who did not understand mathematics. Indeed he never learnt any geometry, but this did not prevent his election in later life to the Royal Society. He could also draw a human likeness beautifully, and as we have seen was an outstanding technical illustrator. He was introduced to a series of sympathetic landowners, who were impressed by his amiability and modesty, coupled with great talent and humble origins. These good introductions provided helpful connections and, combined with the generosity of kind employers, his life progressed. Some of his lectures were turned into publications. His book *Astronomy Explained upon Sir Isaac Newton's Principles* came out in 1756 and was an immediate best-seller. It enjoyed thirteen editions over fifty-six years and the demand for successive reprints did not cease for a further ten years. George III often summoned him to interviews to discuss mechanics.

Ferguson's lectures ranged in title; of matter and its properties, of central forces, on mechanical powers, mills, cranes, wheel-carriages, pile drivers, hydraulic machines, pneumatics, electricity, optics, globes, astronomy, armillary spheres and finally sundials. No doubt it became fashionable for the well-off to attend them, rather like going to the opera today. As with all Royal Society lectures, women were encouraged to attend. His talks were all most lucid and are still eminently readable for their clarity. One contemporary lady commented on an astronomical publication that 'a child of ten years old may understand it perfectly from one end to the other'. All his publications were accompanied by beautiful and explicit illustrations, which were, of course, his own work.

Illustrations of the principles of sundial making made their first appearance in England in the seventeenth century when the subject became part of the curriculum of grammar schools, but James Ferguson's illustrations have never been bettered. All modern illustrations of dials owe something to his draughtsmanship. He claimed to know no geometry. He certainly was ignorant of trigonometry, and the drawings rely entirely on their visual explanation of the principles involved. No mathematics is required. This is their strength. His talents as an illustrator are shown on pages 73, 84 and 117, and the latter engravings demonstrate the relationship between the armillary sphere, the earth, and the equatorial, vertical, horizontal and polar dials. His remarkable drawings provide a fitting close to an age which believed that science was open to all and that reason could solve everything.

The double horizontal dial at Staunton Harold. This is one of the finest garden sundials ever made and is a marvel of English instrument making. It was constructed in about 1685 by the great mathematical instrument maker Henry Wynne. It was probably commissioned by Sir Robert Shirley, later 1st Earl Ferrers and until the 1950s stood in the garden of the family home, Staunton Harold in Leicestershire. It has to be large to carry the wealth of its superlative engraving and measures 30 inches in diameter. The original dial is greatly corroded, but a recent study by John Davis and Michael Lowne explains it. The illustration shows that the gnomon has a vertical edge as well as the usual sloping one. Readings are taken from the shadows of both. The owner now keeps the dial indoors but, following the study, a reproduction of the dial has been made by Tony Moss.

On the large band near the edge are the Roman numerals of a conventional sundial read from the shadow of the sloping gnomon. But there is another horizontal dial. The central area of the plate is dominated by a stereographic projection of the heavenly sphere. The whole of the sky above is projected down in miniature on to the horizontal plate, with the positions the sun occupies at different hours and on different dates shown by the intersections of the many curves. You can see the hours in Roman numerals at top and bottom, and the date curves and dates to left and right. The vertical part of the gnomon is used to read this part of the dial. It will cast a shadow across the curves. You need to know the date of your reading. The point where the shadow intersects the curve for that date is noted and the time is read by judging the time curve that meets

at that same point. Outside the stereographic projection are compass points read against the shadow of the vertical edge to show the direction or azimuth of the sun. The dial works by knowing the direction the sun occupies at any time for a given date. It is therefore known as an azimuth dial, and the observant reader will notice that it is closely related to the analemmatic dial.

Sundials can be used to tell the rough time by moonlight but the moon moves erratically and at different speeds to the sun, so moon time needs adjustment to give sun time. There are many tables engraved on the plate. One of them enables the calculation of the age of the moon – the number of days that have elapsed since the new moon, never more than thirty. Knowing this, the time is read against one of the moon rings appropriate for its age – the rings are outside the compass circle.

The many other scales and tables include measurement of the sun's altitude, calculation of the date of Easter, calculation of the day of the week on which any date falls, the telling of time by the position

of certain stars, and telling the time at thirty-four different places in the world. (These include obvious choices like New York, Paris and Hamburg, but also less obvious places like 'Balasore in ye Mogulls Kingdom'.) Finally, the top edge of the gnomon has a scale giving the Equation of Time against date. This may be the earliest known dial to contain such information.

The Shirley family has enjoyed a long, illustrious, and indeed eventful, history. The great grandfather of the 1st Earl was an ardent Royalist and was imprisoned in the Tower of London by Oliver Cromwell, where he died, some said of poison. The 4th Earl, though able, was unstable and murdered his steward; he was hanged with a silken cord, as was the privilege of condemned noblemen.

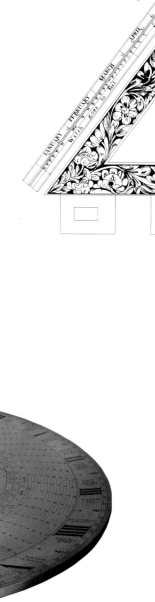

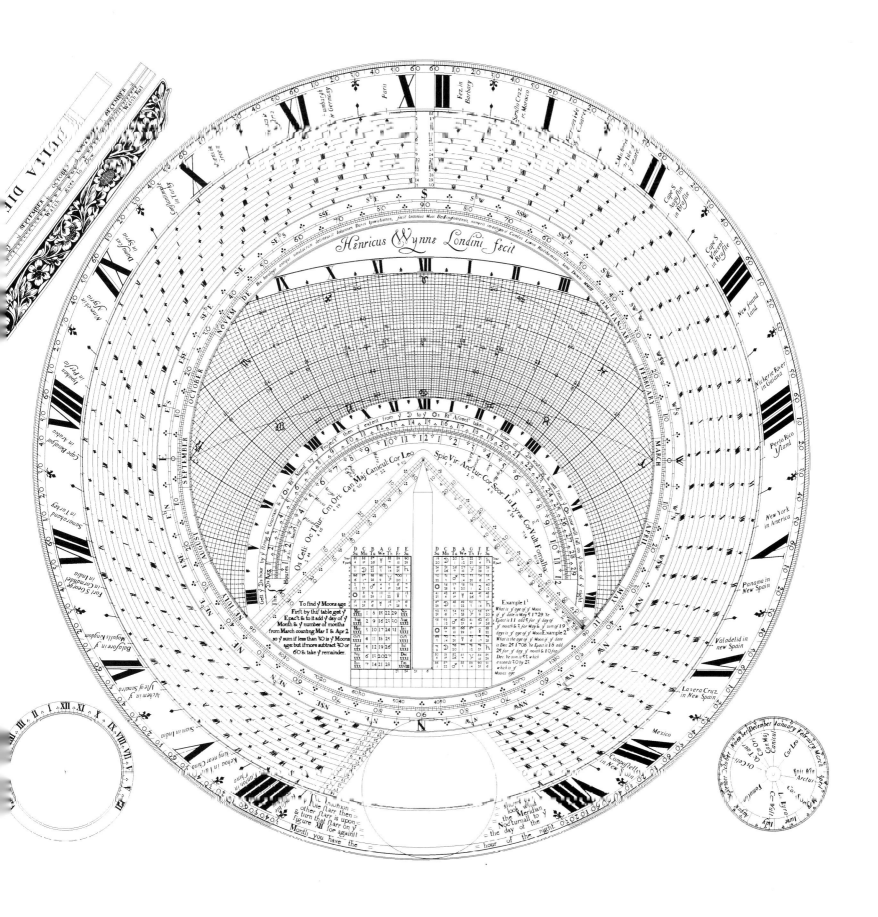

Universal reclining dial, late eighteenth century, by I. Sisson, brass, 8¾ inches square. This is a good example of the kind of dial that was to be found in the houses of wealthy people, and was used to set the household clocks. It demonstrates the quality of the engraving of English craftsmen of the time. It is basically a horizontal dial which can be adjusted for different latitudes. All horizontal dials can be used at any latitude provided they are tilted so that the angle of the sloping edge of the gnomon with horizontal is equal to the latitude. This dial was made to work on this principle. It was made by one of the Sisson family, whose workshop was at the Sphere, corner of Beaufort Buildings, Strand, London. It is provided with spirit levels and levelling screws, and inside the numeral ring is an Equation of Time table. The arms are those of the Graham of Killearn family, kinsmen of the 2nd Earl of Montrose.

The photographer has adjusted the dial face so that the edge of the gnomon is almost horizontal – set for use near the equator!

RIGHT Engravings from *Lectures on Select Subjects* by James Ferguson (London, 1773). Both the horizontal and vertical dials derive from the earth, as shown in the middle of the armillary sphere (above, left and right). The earth is tilted according to the latitude of the dial's location, in this drawing to 51.5°, that of London. Imaginary lines of longitude, b, c, d, e, f and g, each spaced 15° apart, strike the circular ring ABCD at points to which hour lines emanating from the centre E are drawn for the horizontal dial. The gnomon EP is parallel with the earth's axis. In a glance a similar analysis can be seen for the vertical dial.

BELOW The 'universal dialling cylinder' was a visual invention by Ferguson to explain how various sundials are derived. The imaginary cylinder is tilted towards the Pole Star so that its axis EG is parallel with that of the earth, and its outside surface is divided into twenty-four divisions, each spaced 15° apart. The explanation starts with the equatorial dial at B, then follows the vertical at C, the horizontal at F, and finally the polar dial at AD, which is very similar to the equatorial.

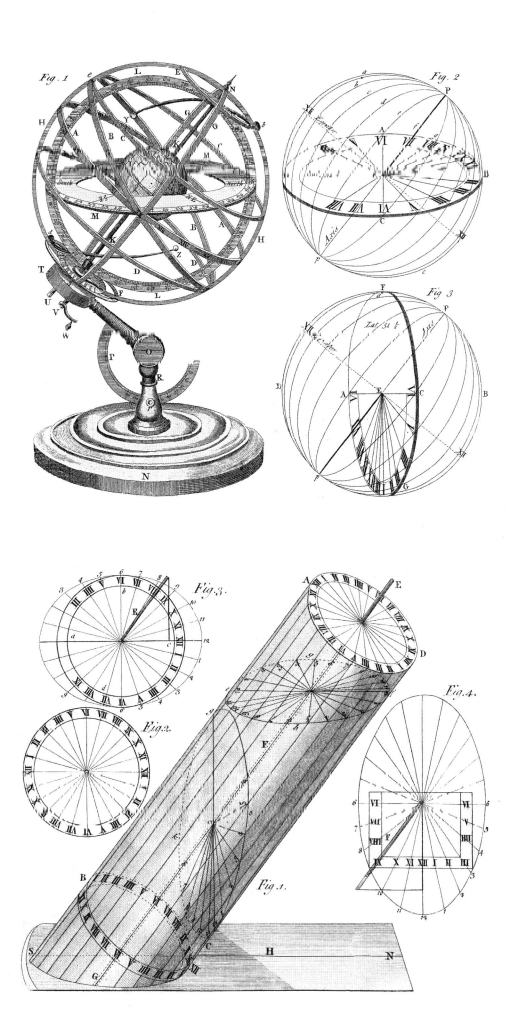

10

SYMBOLS

One of the mysteries in the history of sundials is the enormous interest the subject generated in Scotland in the seventeenth century. All over the country the landed gentry decorated their gardens with spectacular carvings in stone. Commissioned by landowners and the titled, they became increasingly elaborate. Sometimes they were carved on obelisks, sometimes on lecterns; recesses and cylinders were gouged out or curved into shapes of symbolic significance, like hearts for love, globes and stars for the Universe; flat faces were sloped or at angles, and always the dials were complex. Sometimes the workmanship became highly exuberant as at Glamis Castle, where the dial has eighty-four different faces.

These dials are polyhedral. Part of their charm was to demonstrate the skill of their makers in being able to carve on to a block of stone differing projections at different angles, all of which indicated the same time from their different gnomons. By the same token the owner, no doubt a connoisseur of scientific bent, might impress his friends with his understanding of the geometry involved, while the intellectual climate of Calvinist Scotland favoured the study of science and despised decoration for its own sake, seeking function as well as form. Yet their engravings were too small for them to have been accurate enough to set clocks and so it seems unlikely that they were made primarily as timekeepers.

There are some parallels with designs of this date in other countries but Scotland undoubtedly possesses Europe's finest collection of dials in this style. Quite why the fashion developed in Scotland is uncertain. Much earlier, the Renaissance garden had evolved, originally in Italy. It reached England in the sixteenth century, by the end of which a period of prosperity and stable government had developed in Scotland to an extent that enabled the construction of unfortified and open mansion houses with pleasure gardens. One aspect of the Renaissance garden is as an area for the demonstration of the sciences, and this included the incorporation of sundials. We might conjecture that this fashion persisted in Scotland, but all these conditions existed in England and other European countries as well, and none of them explain why it was

Some examples of the great polyhedral dials of Scotland, from *The Castellated and Domestic Architecture in Scotland from the Twelfth to the Eighteenth Century* by MacGibbon and Ross (1887).

Top to bottom, left to right: Drummond Castle, Invermay, Glamis Castle, Neidpath, Woodhouselee, Mount Stewart, Mount Melville, Melville House, Leven, Inch House.

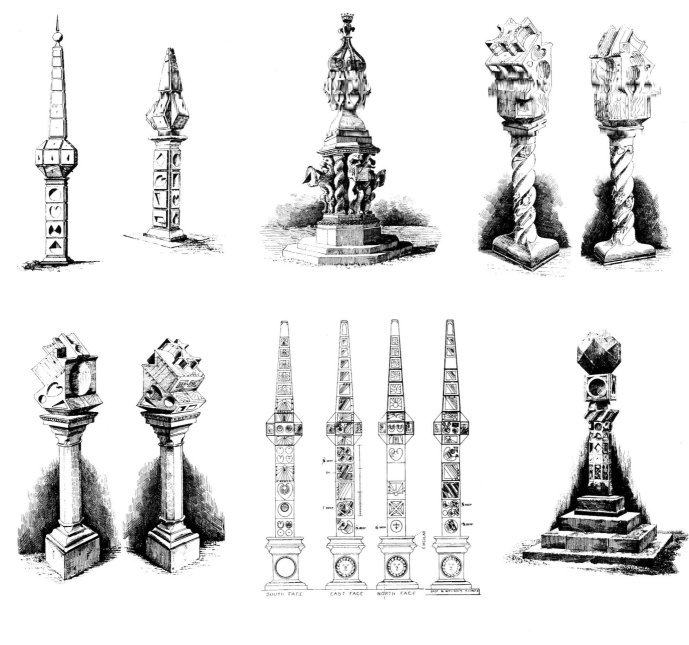

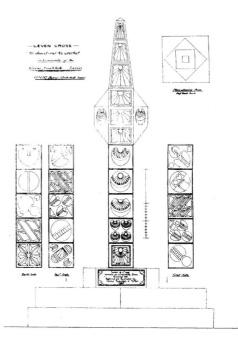

ABOVE Modern polyhedral obelisk dial, by Gerald Laing, with calculations by Ken Mackay, Portland stone, 6 feet high. Commissioned in 2000 by the late Sir Paul Getty. In the grand Scottish tradition it has 32 dials and stands outside Sir Paul's cricket pavilion. Some of the dials show the time in the major cricket zones of the world.

ABOVE RIGHT *Tristan's Sail Wind Shadow*, by Ian Hamilton Finlay, Little Sparta, Scotland. As he lay dying, Tristan sent for Isolde to come and asked her to have a white sail if she was on board or a black one if she was not. Tristan's jealous wife lied to him, saying that the sail was black, and he died before his true love could reach him. The gnomon is the sail, which casts a shadow, and in the shadow of the wind sails are helpless, as were Tristan and Isolde in their sadness.

HORAS NON NUMERO NISI SERENAS
I show only the happy hours

in the latter country that the dials flourished. One theory involves the origin of Freemasonry, which developed earlier than elsewhere north of the English border. It may be that some of the honorary masons who were gentlemen, and certainly unable to carve stone, and who were more likely people with scientific interests, formed friendships with their craftsman colleagues in the lodges. This in turn may have stimulated the fashion in these dials, which were expensive items to make, and required the patronage of the rich as well as the talents of the artisan. It is not clear which of the two designed them. There is no doubt that masons at the top of their profession were capable of designing as well as executing the work. However, as it was also customary at the time for rich gentlemen to design their own houses, the same may have applied to sundials. Some 300 of these polyhedral dials survive. There were certainly more in their day but fashions wax and wane and they disappeared as quickly and as surprisingly as they appeared, for most of them were built over a period of no more than a hundred years.

The sundial continued to be used through the eighteenth century into the nineteenth. As we have seen, there are many fine examples. By the nineteenth century, the finest dials were always for use indoors, often on library tables of great houses, by which the clocks could be regulated. By now, the clock was more accurate than the dial, but for accuracy pendulum clocks and spring watches still needed adjustment fairly frequently. In the nineteenth century came the telegraph and in the early twentieth century the radio, or wireless as it was called. The famous time pips of Greenwich commenced in 1924, and from then, for any practical purpose, the sundial was dead.

So they began to decay. Gnomons became bent or broken, or lost altogether. By the middle of the twentieth century very few new dials of quality were being made, although there were some significant exceptions, for example the Moore dial in Chicago. The portable dial was soundly dead, probably for ever, and certainly in Great Britain those dials that were made for gardens were often cheap items of cast brass, not accurately calculated, intended to sit on a home-made pedestal or one of reconstituted stone. The dial had gone downmarket and was more often than not a companion to the garden gnome. 'Carpe diem', which may be best translated as 'Harvest the day', became a widely used and almost vulgar dial cliché. It comes from an Ode by the Roman poet Horace: 'Be wise, strain the wine and cut back long hope into a small space. While we speak, envious time will have flown past. Harvest the day and leave as little as possible to tomorrow.' It must have been sad for anyone who knew the original Latin to see those debased objects in our garden centres. Happily, towards the end of the twentieth century a reawakening of interest in sundials occurred, modified by the fact that they were no longer items of practical use. The British Sundial Society, now with a substantial and growing membership, was formed in 1989 by some enthusiasts who were astonished to discover that there were others in the world who shared their interest. Well-developed societies also exist in the USA, in the rest of Europe and around the world.

At the same time, there has been a renewal of interest in the design of dials. They are now made by craftsmen who specialize in the subject, and are often commissioned for special occasions as a reminder of some happy or sad event, symbolic of the passing of time. There has often been a symbolic intent in sundial makers and this motive has increased with the demise of the sundial as a practical tool. At the same time the designs have often become abstracted. The inventions of Ono Yukio, Arata Isozaki, Denis Savoie and Santiago Calatrava are more about sculpture and architecture than timekeeping. Ian Hamilton Finlay has, at Little Sparta in Scotland, used the gnomon of a sundial as a visual metaphor for the sail of a boat. Ben Jones in England and Kate Pond in the USA have created beautiful abstractions, which transform the sundial into a sculptural metaphor on the passage of time. Modern sundials are often made to symbolize some important event – a jubilee or the passing of a centenary.

Yet memorial dials are of long standing as well. Lady Anne Clifford was a magnate of the north and died one of the richest landowners in England. She was also greatly loved for her goodness and decency, but much of her life had been a struggle as a result of the demanding legal actions she was forced to pursue to secure her inheritance. In this she enjoyed the affectionate support of her mother and she reciprocated this loyalty also with deep love.

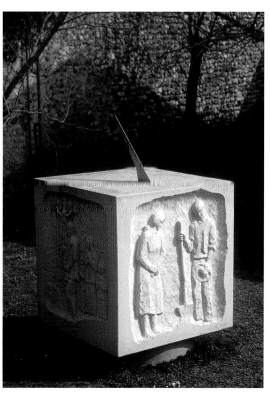

LEFT AND BELOW Sundial Bridge, Turtle Bay, Redding, California, by Santiago Calatrava, steel, glass, granite and concrete. Opened July 2004. The renowned Spanish architect has built bridges, airports, rail terminals and stadiums throughout the world. This bridge spans the Sacramento River and links the north and south campuses of the Turtle Bay Exploration Park.

TEMPUS VINCIT OMNIA

Time conquers all

TOP LEFT *Himeguri*, 1998, by Kate Pond, Mitsubishi Sports Garden, Sendai, Japan, corten 'weathering' steel and Date kan muri boulders, 8 feet × 24 feet × 75 feet. The illustration shows the midsummer solstice.

TOP RIGHT *Sunfix*, 1998, by Kate Pond, Highgate Springs, Vermont, USA, corten 'weathering' steel and black Vermont marble, 8½ × 8½ × 9½ feet. The photograph shows an image of the sun at the autumn equinox aligned with the ellipse of black marble.

BELOW *Zigzag*, 1991, by Kate Pond. Steel maquette for Colby Curtis Museum, Stamstead, Quebec, Canada. It is equinox when the ponted shadow reaches the horizontal leg.

Their last tearful farewell on horseback took place near to a family property, Brougham Castle, a month before her mother died. Many years later, in 1654, Lady Anne erected on this spot a beautiful dial on a column in memory. It is near to what was then no doubt a quiet lane, but which is now one of the busiest roads in north England. Filial affection is to be seen in a simple and charming sundial sculpture carved by Alfeo Mizzau for the garden of the family home in north Italy. It depicts images in the life of his father and mother, who were peasant farmers, and bears an inscription in local dialect extolling the skills of his mother. She spent so much time labouring outside in all seasons when the sun was low as well as high that she could, as some human sundial, and following Isaac Newton, tell the time without consulting a clock or watch.

LEFT *La Nef Solaire*, Languedoc, France. Called 'the solar nave' because it has the appearance of both a church or a boat with sails, this remarkable sculpture was designed in 1993 by Denis Savoie, President of the Sundial Commission of the French Astronomical Society. It is to be seen along the autoroute A9 between Orange and Nîmes. It is built in concrete and weighs 240 tons with an overall height of 55 feet. It is a masterpiece of construction by the sculptor Odile Mir and the engineer Robert Queudot because of the considerable difficulties in aligning such a massive structure with accuracy.

RIGHT Kentucky Vietnam Veterans Memorial. This giant sundial, which measures 89 × 71 feet, was designed in 1987 by Helm Roberts of Lexington, Kentucky, and was a considerable feat of design and construction. To allow for water drainage the plaza is sloped 1:50; the dial is therefore strictly not horizontal, but reclining.

The tip of the sloping gnomon tells the date, and at the death of each Kentuckian, the shadow of the gnomon touches the name in brief acknowledgment of the debt owed to him by his country.

FAR RIGHT Memorial sundial, Piazzale San Giusto, Trieste, 1918. The inscription commemorates a long-forgotten advance of the Italian army against the Austrian forces in the First World War, in the words of the Commander in Chief: 'Our forces occupied Trent at 3pm and set out for Trieste at 4pm. The Tricolor flies over the castle of Buon Consiglio and over the tower of S. Giusto.'

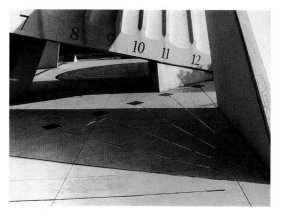

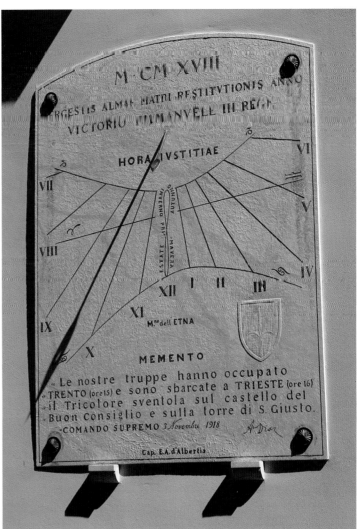

UMBRA SICUT HOMINIS VITA

The life of man is like a shadow

Sundials are used to remember tragedy. The Kentucky Vietnam Veterans Memorial at Lexington KY is a plane sundial in a gigantic plaza paved in massive slabs of granite. It tells time as well as date and the names of all who died are engraved in the months of their passing. On a similar but perhaps slightly happier note for some, there is, in northern Italy, a range of fine sundials made in different towns after the First World War by a senior officer who was an expert. They are in memory of the battles and victories, of great significance in the hearts and minds of the participants, but long since forgotten by most.

So sundials are often about recollection, sad or happy, as are their mottoes. Mottoes are always found on coats of arms and sometimes on gravestones. They have often been used on sundials to recall some universal truth or saying. They will have been seen scattered in this book at odd moments when they seemed to have relevance to the text. The passing of time and the imminence of death are perhaps the most powerful messages to be found. If you look Holbein's painting *The Ambassadors* (see page 78) at a sharp angle from the bottom right hand side, you see the image of a skull, which in normal viewing is made invisible by distortion. Rossetti used sundials in a number of paintings, as images of death, nostalgia and romance. In *Beata Beatrix* the sundial's shadow falls at nine, the number that Dante associated with his lover, Beatrice, both throughout her life and at the hour of her death.

This is not a book about every aspect or design of sundial. It says nothing about astrolabes or quadrants, both of which can be used to tell the time, nor does it mention several other forms of dial which are of lesser importance. Nor, perhaps thankfully, does it include much mathematics. It is written for those who may be interested in the central story of measuring time without the use of clocks. Two emperors, two popes, a maharajah and the founding president of the United States of America have been sufficiently interested to have sundials made. Shakespeare and Chaucer referred to them; Newton, Wren and Dürer designed them. So anyone who has read this book this far is in good company. Sometimes the story is surprising and about brilliant and eccentric people. Always there is a thread of art. As to science, the terms used are largely simplified. Anyone who delves into this subject more deeply will discover that what I call the equatorial dial is by purists usually named the equinoctial dial – the latter a confusing term for the dial has less to do with the equinoxes than with the equator.

In the eighteenth century James Harrison invented the ship's chronometer, a device so accurate that in 1764 it lost only 39 seconds on a journey of 46 weeks from England to Jamaica, notwithstanding the motions of the ship, the storms and changes in temperature and humidity. His land-based regulator clocks could tell time to one tenth of a second. Caesium fountain clocks now tell time to one second in 15 million years and physicists at the National Physical Laboratory in London are working on the ion trap, a time-measuring standard that may enable scientists to measure time to one second in 15 billion years. As this is the age of the universe itself, it may also be the age of all time. But of course we do not know, for as some have speculated, there may have been other existences before the big bang gave birth to our universe.

Such accuracy suggests that sundials, already less accurate than clocks 250 years ago, should be obsolete, but they are not. In the year 2004 the first interplanetary sundial was sent to Mars, largely because there is a renewal of interest in the subject. Sundials are arguably the oldest of scientific instruments. Their beauty is often a reflection of great craftsmanship, as well as a design statement which is highly ordered and rational, and to an extent out of the hands of the designer, who must always find a solution that both obeys the rules and satisfies the eye. There is a particular symbolism in an object that does something helpful but requires no power and performs indefinitely. This must signify endurance and permanence. Hence the great constructions of Augustus and the Maharajah, but there is no permanence even in masonry, only the pretence of it. Most of Jai Singh's observatories have gone completely and the Orologium is buried under two millennia of ancient Rome's decline. Ulugh Beg's observatory and Uranibourg are ruins, masonry metaphors for the passing of time. Endurance leads to constancy, remembrance and love, as in the war memorials, the Scottish dials with their heart carvings, and Lady Anne's pillar. We carve inscriptions and mottoes on objects to remember people, events, wise sayings and anything that should not be forgotten and can be stated in a few words, as with religious sentiment or the anticipation of death. Sundials have always been suitable for mottoes, because they are symbols of permanence. Of all their symbols perhaps the greatest is what has troubled man since he was first capable of thought and still troubles him today; and that is the mystery of the passage of time itself.

ABOVE Dial sculpture, 1993, by Ben Jones, limestone, 22 inches high

RIGHT Mars probe sundial, 2004. A dial like this was sent on each of the two Mars rovers in 2004. The main purpose of the dials is found in the grey scale rings, which are used to calibrate the unmanned spacecrafts' cameras for colour values. The rather unusual gnomons and dials were added in order to inspire schoolchildren of the twenty-first century with a love of sundials.

The dials carry the word 'Mars' in seventeen languages, including Mayan and Sumerian, just in case the rovers were, through some strange accident, to return to earth via a wormhole and end up in 3000 BC.

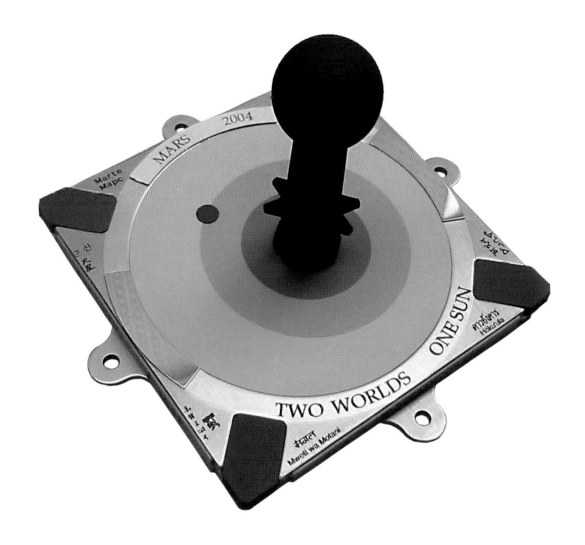

TEMPUS FUGIT *Time flies*

APPENDIX
MOSTLY ON GEOMETRY AND THEORY

ANGLE OF STYLE

As was stated on page 59, the first of the basic rules of the equatorial dial states that the angle of the style with horizontal will be equal to the latitude of the place. This rule holds true for all dials with slanting gnomons. Look at Diagram A. You will see a small horizontal dial drawn in profile at a place with latitude (ϕ). The line of the style must be parallel with the axis of the earth.

HORIZONTAL DIAL

Diagram C shows how to draw a horizontal sundial without any maths but sundials are much easier to calculate than they seem and may be understood by anyone who remembers or who is happy to be reminded of the basic rules of trigonometry. In any right-angled triangle the relationships between the sides and the angles are as follows:

$$b \div a = \sin B$$
$$c \div a = \cos B$$
$$b \div c = \tan B$$

You may be surprised to know that you need remember no more mathematics than the above to understand the horizontal dial and only a little more for the vertical dial. Everything derives from the equatorial dial. Diagrams B and C show the relationship between an equatorial and a horizontal dial. The style of the gnomon of the equatorial dial (marked in blue) is tilted to the latitude of the site, and therefore follows the basic rules, being parallel with the axis of the earth. The style of the horizontal dial follows this line and the projections of the hour lines from the equatorial disc are shown. Imagine this folded out, that is, the equatorial disc folded forward along the line where it touches the horizontal, and the style folded to the right along the line of noon. By unfolding the design it is possible to draw it with pencil, ruler, set square, compass and protractor.

Diagram C shows B unfolded. The hour divisions (in red) on the equatorial ring are, according to the rules, each 15°. These are known as hour angles, which are measured from noon, both into the afternoon and backwards into the morning. Thus an hour angle of 15° corresponds to 1pm and 11am, 30° to 2pm and 10am. Noon is zero. It follows that an hour angle of 22.5° (15 × 1½) corresponds to 1.30pm and 10.30am. One can go on. Every 3.75° division (15 ÷ 4) corresponds to 15 minutes of hour angle and every 1.25° (15 ÷ 12) corresponds with 5 minutes and so forth. So, for example, if the time is 1.35pm or 10.25am the hour angle is 23.75° (15 + [15 × ^{35}/$_{60}$]). You can make as many sub-divisions as you want, for single minutes if that is your wish.

To draw this is straightforward. From point O, draw the blue vertical line to any convenient length a. This becomes the 12 o'clock line. From O mark off the latitude ϕ, which we know, and from E drop a perpendicular b to G. Project a to C so that b' = b. From C draw the red semi-circle and mark off 15° divisions as shown. From point E draw a horizontal line to right and left as shown by the line 16 to 8. Project the red 15° divisions to line 16 to 8 and you have the hour points to be joined to O to complete the basic layout.

Trigonometry and Proof

It is better to calculate all this. The broken red lines are drawn to show *any* hour angle **h** on the equatorial disc and *any* time line on the horizontal dial. Let us say that such a line makes an angle of **X** with noon and then calculate **X** in terms of **h** and ϕ.

$$\tan X = DE \div a = (DE \times \sin \phi) \div b = (DE \times \sin \phi) \div b'$$
$$= \sin \phi \times \tan h.$$
$$\tan X = \sin \phi \times \tan h.$$

This is the most basic formula in all sundial mathematics and with it one can calculate for a horizontal dial the angle **X** that any time line of hour angle **h** makes with noon, where ϕ the latitude is known.

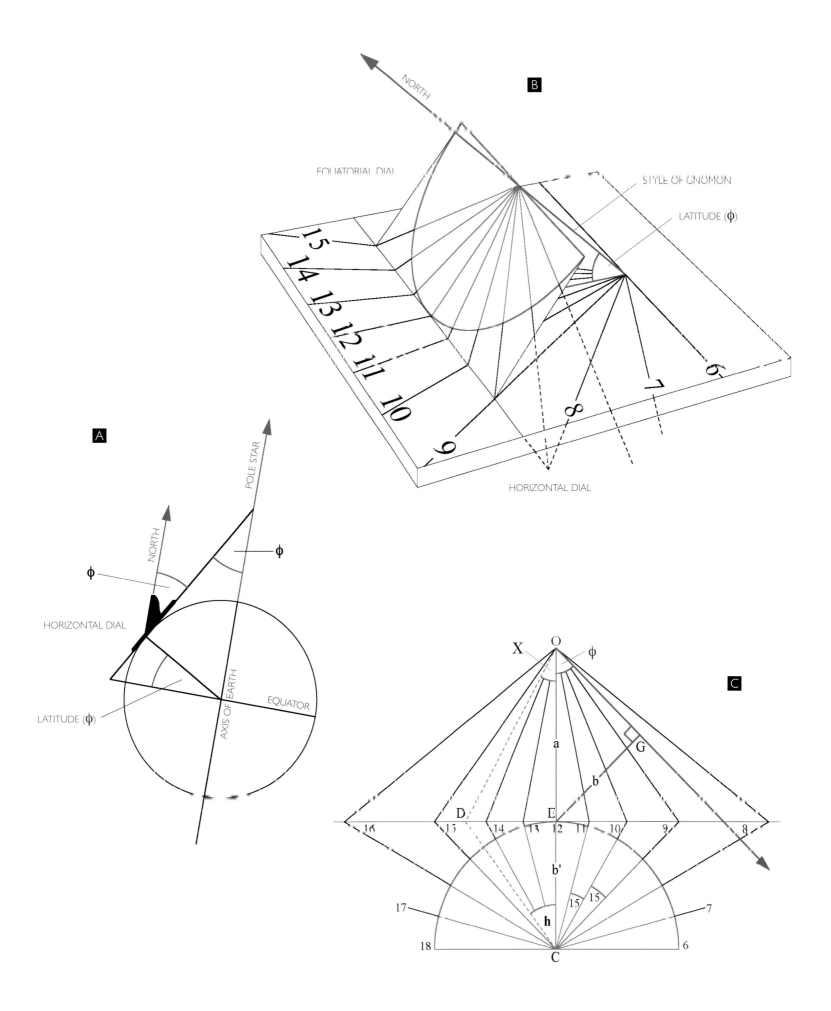

B

NORTH

EQUATORIAL DIAL

STYLE OF GNOMON

LATITUDE (φ)

15
14
13
12
11
10
9
8
7
6

HORIZONTAL DIAL

A

POLE STAR

NORTH

φ

φ

HORIZONTAL DIAL

LATITUDE (φ)

AXIS OF EARTH

EQUATOR

X

O

φ

C

a

b

G

D

E

16 15 14 13 12 11 10 9 8

b'

17

15 15

7

h

18

6

C

VERTICAL DIAL

Those who have followed the explanation for making a horizontal dial will have suffered little pain. It is only slightly harder for a vertical dial but the description will be more easily understood with several diagrams, which move from one stage to the next.

Diagram D shows how the vertical dial relates to the horizontal dial. Only four hour lines are shown. Several other features should be noted. The wall is not aligned precisely west–east but is turned through an angle of **d°**. It is said to decline **d°** to the west. Some walls, of course, decline to the east, in which case the geometry will be understood if you imagine a tracing of this drawing which is turned back to front. The style of the gnomon in blue for the horizontal dial is in its correct position on the noon line and the style of the vertical dial follows it precisely. The time lines are drawn from the point at which the vertical style strikes the wall. They coincide with the time lines of the horizontal dial at the point of intersection of the two planes. The vertical style is therefore an extension to the wall of the horizontal style. Also note that the vertical style makes an angle **SH** (style height) with the wall. Its plane is perpendicular to the wall (see Diagram E) and its base makes an angle **SD** (style distance) with the noon line.

Diagram E is a tracing of D but with all the hour lines except 12 removed. Note that on *all* vertical dials the noon line is always vertical. Remember that the plane of the gnomon, i.e. the blue triangle in E , should be perpendicular to the wall. Note again **SH**, **SD**, **d** and **ϕ**. A three-dimensional shape is made up of four triangles with the letters a, b, c, e, f and g, referring to the six sides of the shape and their lengths. First one needs to know values for the angles **SD** and **SH**.

Trigonometry and Proof

One could work the geometry out and draw it to measure them, but anyone making a vertical dial would be foolish not to use a calculator with trigonometry, or a computer, for at least part of the work. For this reason only the maths is given to ascertain **SD** and **SH**, which shows that

$$\tan \mathbf{SD} = b \div a = b \div (c \times \tan \boldsymbol{\phi}) = \sin \mathbf{d} \div \tan \boldsymbol{\phi}$$
$$\text{and}$$
$$\sin \mathbf{SH} = e \div f = (e \div c) \times (c \div f) = \cos \mathbf{d} \times \cos \boldsymbol{\phi}.$$

Diagram F shows how to draw the dial with or without maths, and how to calculate the hour lines with a method very similar to that for a horizontal dial. First, draw a vertical line of any

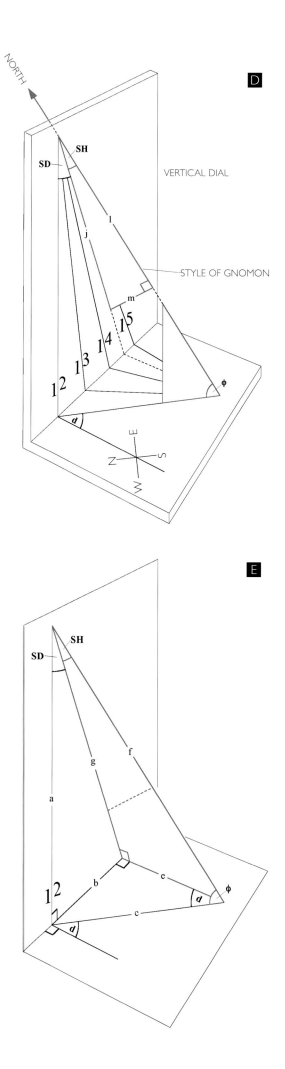

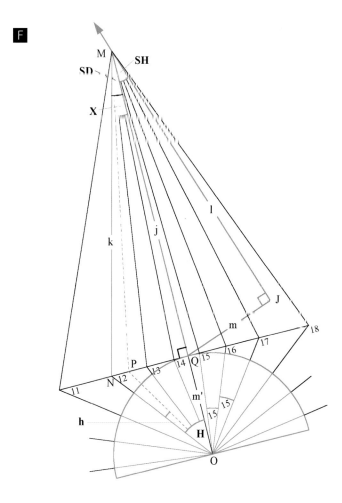

F

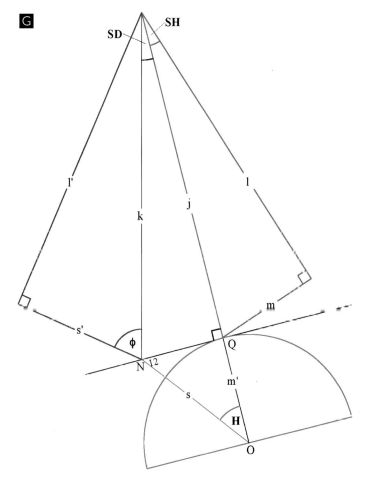

G

length, here described as k. This will be the 12 line. Now draw j of any length and at angle **SD** to k. From any point Q draw a perpendicular to strike k at N and extend both ways to make the line marked 11 to 18. Draw a line at angle **SH** to j and from Q drop a perpendicular to it to make m and l. This blue triangle is the gnomon folded flat. Extend j to O to make m' of length equal to m and draw the red semi-circle centred at O. Join N to O, and from this line mark off 15° radii and extend them to the 11 to 18 line to make the 11 to 18 hour points and the hour lines. Ignore for the moment the broken red lines, defining **X** and **h**, and also ignore **H**.

The above can be calculated but a further diagram is necessary for the proof. Diagram G is a simplified tracing of Diagram F. The hour lines except 12 have been removed. There are some additions in brown: s' is drawn at angle ϕ to k such that s' is equal in length to s, the red line NO; l', which is equal in length to l, is drawn perpendicular to s'. It can be imagined that Diagrams F and G contain a three-dimensional shape folded flat, rather like Diagram C for the horizontal dial. The blue and brown triangles fold up with l and l' forming one edge, and the semi-circle is folded up along the 11 to 18 line to meet the triangles; s and s' form one edge, as do m and m'.

Note from Diagram F the broken red line, which is any hour angle **h** and which makes any angle **X** with j the base of the gnomon. Note that the angle between m' and NO is marked of value **H**. This is defined as the angle the noon line makes with the base of the gnomon in the plane of the equatorial disc.

From Diagram G:
$\sin \mathbf{H} = NQ \div s = k \times \sin \mathbf{SD} \div s = k \times \sin \mathbf{SD} \div s' = \sin \mathbf{SD} \div \cos \phi$.

From Diagram F:
$\tan \mathbf{X} = PQ \div j = m \div j \times PQ \div m' = \sin \mathbf{SH} \times \tan(\mathbf{H} - \mathbf{h})$

The formulae which have been proved are:
(1) $\tan \mathbf{SD} = \sin \mathbf{d} \div \tan \phi$
(2) $\sin \mathbf{SH} = \cos \mathbf{d} \times \cos \phi$
(3) $\sin \mathbf{H} = \sin \mathbf{SD} \div \cos \phi$
(4) $\tan \mathbf{X} = \sin \mathbf{SH} \times \tan(\mathbf{H} - \mathbf{h})$

It follows that (1), (2) and (3) are used first to calculate **SD**, **SH** and **H**. When these values, which do not vary, are known, equation (4) can be used to calculate values of **X** for any value of **h**. All this assumes that the value d for the declination of the wall is known. How to measure this is shown later.

ZODIAC AND DATE

A little has been written earlier about the zodiacal curves and date. Both a graphical and mathematical solution to drawing them are given below. The explanations use declination values for the zodiac curves, but obviously other values for actual dates can be used instead. Declination varies only very slightly from year to year, and mean values should be used for sundials. Mean values of declination for all days of the year are given in many of the books in the bibliography and in British Sundial Society (BSS) and North American Sundial Society (NASS) publications. Visit their web sites. Actual values, for a date in a particular year, can be found in publications such as *Whitaker's Almanack*, or on the Internet.

The Trigon

First you should understand the trigon (see Diagrams H, H' and I. H, the vertical dial, is derived from D). The trigon is an instrument that was used in the past for setting out dials without calculation. It was clamped to the end of the style. The divisions of the zodiac are shown on the enlarged drawing of it at H'. From the illustration on page 31 we know that the winter solstice declination (the start of Capricorn) is −23.45°, and that the summer solstice declination (the start of Cancer) is +23.45°. On the trigon the angle that the Capricorn mark makes with the equinoctial line is 23.45° – the same for Cancer, the other side of the equinoctial line. All the zodiac periods are marked off in angles according to their declination values, as shown in Diagram H'. In sundial terminology, δ is normally used for values of solar declination.

The trigon was turned on the gnomon rod, and was equipped with a strong thread that could be stretched from the gnomon tip. It was turned until its plane coincided with a particular time line, and the thread was stretched along the trigon, past the appropriate declination reading on to the face of the dial, to make a mark. When a sufficient number of points had been marked with the thread, they were joined into a curve – or, in the case of the equinox, a straight line. In H the trigon is shown clamped to the style with the Cancer and Capricorn curves and the equinox in their approximate positions.

Graphical

You must first draw all the time lines of your dial. Look at Diagram I, which is derived from F, with much removed. The gnomon in blue is there, as is any time line at angle X in broken red, referred to as X time. The 11 to 18 line is there (with only three time lines: 11, 12 and 18) and has been renamed

'EQUINOX'. The right angle at J is as before, and from J seven broken blue lines are drawn. $\delta 0$ is an extension of m and the others are drawn at the angles to m that they are named with: $\delta + 23.45$ is at 23.45° below $\delta 0$ from J, and $\delta − 23.45$ is at 23.45° above $\delta 0$, and so forth. $\delta + 23.45$ represents the summer solstice because the sun's declination then is +23.45°; likewise $\delta − 23.45$ for the winter solstice.

From the explanation of the trigon, the sun shadow at date $\delta 0$ forms a straight line at right angles to the base of the gnomon, i.e. the line renamed 'EQUINOX', when the shadow of the point J will follow this line. In the earlier example we imagined the trigon working in three dimensions. Now we must imagine it fixed to the end of l at J and, as in Diagrams F and G, the whole gnomon is turned about j to be flat on the paper. Now, instead of using the thread on the trigon, we displace the time lines. The X time line strikes the equinox at P. From the centre at M draw an arc through P to strike $\delta 0$ at P'. The situation is the same as if the trigon had been fixed to J in three dimensions. MP = MP' = the distance, at X time, from M of the equinox line. Points are marked where the broken red line (MP') strikes the broken blue lines, and arcs, centred on M, are drawn across to the X time line. Now, MQ' is the point on MP' which intersects the $\delta − 23.45$ line. So MQ = MQ' = the distance from M at X time at a declination of −23.45°.

Points like Q are drawn for all the declinations (the point R and R' for the summer solstice is also given a letter, but not the others), and for all the times marked on the dial. The points are then joined into curves. All this is immensely laborious and it is much better to do it by calculation.

Mathematical

First, j is fixed by the design and l = j × cos SH, so both j and l are known. Find MP'. MP' = MP = j ÷ cos X. Call angle MP'J by symbol t.

sin t = l ÷ MP' = l ÷ MP = l × cos X ÷ j.
So t = arcsin (l × cos X ÷ j).

Find MQ. Call angle MQ'J by symbol w. If any value of declination is described by symbol δ, then w = δ + t = δ + arcsin (l × cos X ÷ j). From the sine formula:
MQ' ÷ sin (90 − δ) = l ÷ sin w.

$$MQ = MQ' = \frac{l \times \cos \delta}{\text{Sin} \left[\delta + \arcsin \left(\frac{l \times \cos X}{j}\right)\right]}$$

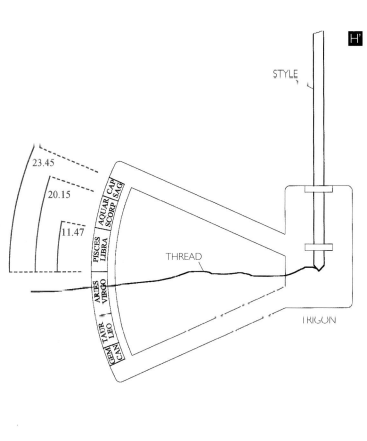

23.45
20.15
11.47

AQUAR CAP
SCORP SAG
PISCES
LIBRA
ARIES
VIRGO
GEM TAUR
CAN LEO

STYLE

THREAD

TRIGON

H'

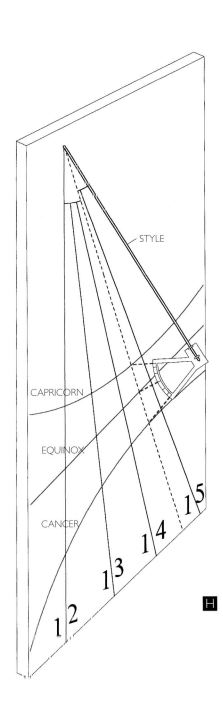

STYLE

CAPRICORN

EQUINOX

CANCER

15
14
13
12
11

H

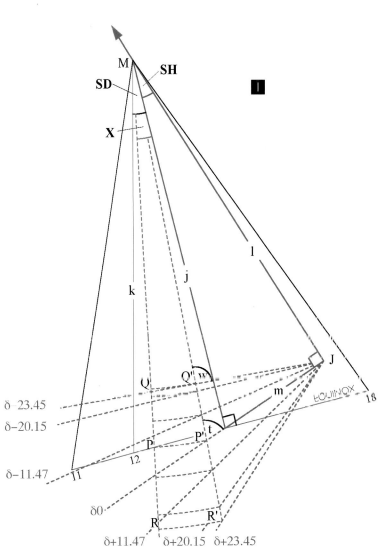

M
SH
SD
X
I

l

j

k

Q
Q' w
m
EQUINOX

δ−23.45
δ−20.15

P'
P' t

12

δ−11.47
11

δ0

R'
R'

δ+11.47 δ+20.15 δ+23.45

18

J

FESTINE LENTE

Make haste slowly

With this formula, you can calculate the distance from M along any time line for different values of δ and **X**.

In this description zodiac date is read from the tip of the style, but in many dials, for example the one at Queens' College (see page 106), there is a small sphere (called a nodus) some way down the style from the shadow of which the zodiac is read. This enables the style to be longer so that its shadow reaches the time divisions marked at the bottom of the dial.

DECLINATION OF WALL

A confusion can easily arise between the declination of a wall, which by convention is usually noted as **d**, and the declination of the sun which is known by δ. For a vertical sundial the value of **d** must be measured. Do not imagine that this can be done easily with a magnetic compass. There are two problems, one of which is often known also by the word declination! Magnetic declination and instrument deviation can cause serious inaccuracies and are not easy to adjust for. Deviation is caused by pieces of metal near the compass, and magnetic declination by the movement of the earth's magnetic poles. In 1850 declination caused an error of 22°, today about 3°.

The best method of measurement is to use the sun. An easy but not so accurate way for those who wish to avoid trigonometry is to make a horizontal sundial and set it up close to the wall on some simple turntable, turn it to the correct time (adjusted from clock time for the Equation of Time and displacement from the local time meridian), and then measure the angle the plane of the gnomon makes with the wall, from which **d** is known.

However the best method is to observe the azimuth of the sun at a known time and use this value to calculate the wall's declination. The azimuth is the direction of the sun, measured as on a compass clockwise from 0° for north. You need a board fixed precisely vertically in the position of the dial with a pointer of metal perpendicular to the board and so horizontal. Its end should be pointed and its length from the point to the board precisely known. Call this **a**. Look at Diagrams J and K. At a known time, the distance **b** is measured. **z** is the azimuth and **d** is the declination. It follows that tan (**d** + **z**) = **b** ÷ **a** and that **d** = arctan (**b** ÷ **a**) – **Z**.

Z is given by the following formula:

$$ \mathbf{Z} = \arctan \left(\frac{\sin \mathbf{h}}{\sin \delta \times \cos \phi - \sin \phi \times \cos \mathbf{h}} \right) $$

This equation is well known in the story of navigation as the Mariner's Formula, for it was in universal use before the invention of global positioning systems. Several values for **d** should be taken and an average made. **h** must, of course, be calculated by adjusting clock time for the Equation of Time and displacement.

LATITUDE AND LONGITUDE

If you cannot always think which is which, remember that latitude goes **a**cross. The best way to discover these values must be from maps or the Internet, in the UK from the Ordnance Survey publications or from www.maporama.com, in the USA from www.geocode.com. Global positioning systems obviously can be used, but such accuracy is not necessary. There are 60 minutes in a degree and 60 seconds in a minute.

Time and Equation of Time

Precise time is obtainable from radio signals and telephone clocks. It is also worth investing in a stopwatch. Daily values for the EoT are not given in this book, but they are available in astronomical tables (ephemerides), in the UK in *Whitaker's Almanack*, in the US in publications by the North American Sundial Society (NASS) at www.sundials.org, on the Internet, and in the books in the bibliography that are starred as of practical value. Approximate values in minutes and half minutes, good enough for the casual reading of a dial, are given on page 136 for every five days of the year. Fast means that the clock will be faster than the dial; slow slower. For example, on 6 January the clock is 6 minutes faster than the dial; on 30 April it is 3 minutes slower. Values for days between given dates can be estimated by interpolation. The EoT is zero on 15 April, 12 June, 1 September and 25 December.

These values can be combined with the displacement of the dial from the time meridian to make composite values. A horizontal dial should be set up by turning it until it reads the correct sun time which when corrected gives the right time against the second hand of a watch. To do this it must be better to refer to an ephemeris for EoT values rather than the table.

Anyone who gets muddled about whether to add or subtract the value should not lose confidence. They are in good company, and there has been no loss of life. Many catastrophic errors of navigation have been made by good people with mistakes of this kind. The only solution is to check again when the brain does not feel jaded. The same goes for the displacement from the time meridian. There are two golden aids to commit to memory:

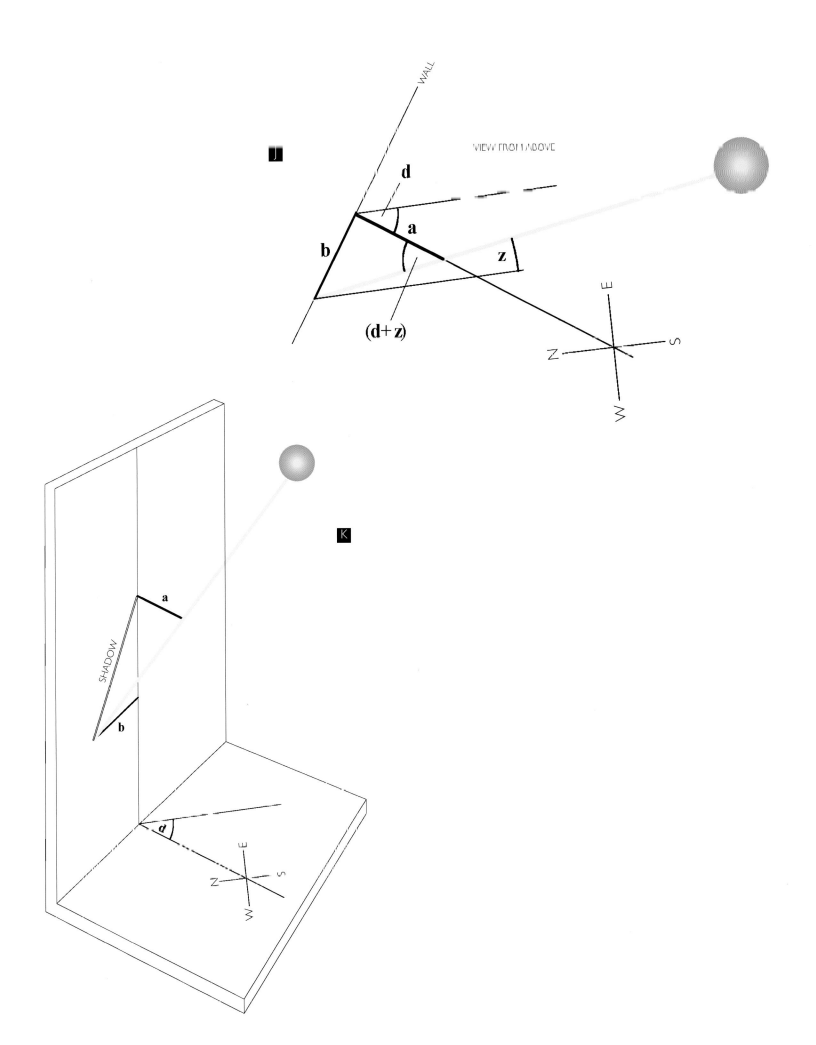

WALL

VIEW FROM ABOVE

J

d

a

b

z

(d+z)

E

N

S

W

SHADOW

a

b

K

d

E

N

S

W

Day	January	February	March	April	May	June
1	fast 3½	fast 13½	fast 12½	fast 4	slow 3	slow 2
6	fast 6	fast 14	fast 11	fast 2½	slow 3½	slow 1½
11	fast 8	fast 14	fast 10	fast 1	slow 3½	slow ½
16	fast 10	fast 14	fast 8½	0	slow 3½	fast 1
21	fast 11½	fast 13½	fast 7	slow 1½	slow 3½	fast 2
26	fast 12½	fast 13	fast 5½	slow 2	slow 3	fast 3
30	fast 13	-	fast 4½	slow 3	slow 2½	fast 4

Day	July	August	September	October	November	December
1	fast 4	fast 6½	0	slow 10½	slow 16½	slow 11
6	fast 5	fast 6	slow 1½	slow 12	slow 16½	slow 9
11	fast 5½	fast 5	slow 3½	slow 13½	slow 16	slow 6½
16	fast 6	fast 4½	slow 5	slow 14½	slow 15	slow 4½
21	fast 6½	fast 3	slow 7	slow 15½	slow 14	slow 2
26	fast 6½	fast 2	slow 8½	slow 16	slow 12½	slow ½
30	fast 6½	fast ½	slow 10	slow 16½	slow 11½	slow 2½

EoT negative: clock fast
West of Meridian: clock ahead

Note: LAT (Local Apparent Time) has been described throughout the book as sun time; LMT (Local Mean Time) means sun time corrected for the Equation of Time but not longitude.

COMPUTER PROGRAMS

It is better to design sundials with trigonometry, rather than with ruler and compass. A calculator with trigonometry is suitable for calculation but the most sensible method, and not difficult if you do not object to trigonometry, is to use computer programs. The really keen will write their own, in Basic or Qbasic, or DeltaCad. The British and North American Sundial Societies publish some information on this subject. All the proofs given in this book are sound, but rather than turn them into computer programs the reader may prefer to use other material.

There is a remarkable program, developed in France by François Blateyron. With it you can design horizontal and vertical dials, pillar and analemmatic dials. There is much other information, which is also very useful. No knowledge of any mathematics is required. It is most user-friendly, and draws fine images of the time lines and date curves. The basic program can be downloaded for free from the Internet, and a more advanced version can be purchased for a modest price. It is available in several languages. Visit www.shadowspro.com.

ACCADEMIA OLIMPICA

Throughout Europe in the latter part of the sixteenth century were founded academies, institutions for exchanging and propagating learning. The Accademia Olimpica, one of the earliest, was chartered in Vicenza in 1555 by a group of twenty-one local scholars, mathematicians, several aristocrats and one artist: the great architect Andrea Palladio. According to its charter its objectives were the cultivation of the arts and sciences, but especially mathematics – 'the true ornament of all who possess noble and virtuous spirits'.

BIBLIOGRAPHY

Andrews, William J H. (ed.) *The Quest for Longitude* (Harvard University, 1996)

British Sundial Society Bulletins, www.sundialsoc.org.uk

Brookes, Alexis, and Stanier, Margaret, *Cambridge Sundials* (Pendragon, Cambridge)

Buchner, Edmund, *Die Sonnenuhr des Augustus* (Verlag Philipp von Zabern, 1982)

Catamo, Mario, *Il Cielo in Basilica – La Meridiana di S. Maria degli Angeli* (Agami, 2002)

The Compendium (North American Sundial Society)

Courtright, Nicola, *The Papacy and the Art of Reform in Sixteenth-Century Rome: Gregory XIII's Tower of the Winds in the Vatican* (Cambridge University, 2003)

★ Cousins, Frank W., *Sundials* (John Baker, London, 1969)

Cowham, Mike, *A Dial in Your Poke* (Cowham, Cambridge, 2004)

Daniel, Christopher St J.H., *Sundials* (Shire Publications, Aylesbury, 1997)

Daniels, Stephen, *Joseph Wright* (Tate Gallery, London, 1999)

Duncan, David Ewing, *The Calendar* (Fourth Estate, London, 1999)

Evans, James, *The History and Practice of Ancient Astronomy* (Oxford University, 1998)

Fantoni, Girolamo, *Orlogi Solari* (Technimedia, Rome, 1988)

Foister, Susan, et al., *Holbein's Ambassadors: Making and Meaning* (National Gallery, London, 1997)

Gatty, Mrs Alfred, *The Book of Sundials* (Bell & Sons, 1900)

Gibbs, Sharon L., *Greek and Roman Sundials* (Yale University, 1976)

Gleick, James, *Isaac Newton* (Fourth Estate, London, 2003)

Gouk, Penelope, *The Ivory Sundials of Nuremberg 1500–1700* (Cambridge University, 1988)

Heilbron, J.L., *The Sun in the Church: Cathedrals as Solar Observatories* (Harvard University, 1999)

Herbert, A.P., *Sundials Old and New* (Methuen, London, 1967)

Higton, Hester, *Sundials at Greenwich: A Catalogue of the Sundials, Nocturnals and Horary Quadrants in the National Maritime Museum, Greenwich* (Oxford University, 2002)

_____, *Sundials: An Illustrated History of Portable Dials* (Philip Wilson, London, 2001)

Janin, Louis, *Le Cadran Solaire de la Mosquée Umayyade à Damas* Vol. 16 (Centaurus, Copenhagen, 1972)

Jardine, Lisa, *On a Grander Scale: The Outstanding Career of Sir Christopher Wren* (HarperCollins, London, 2002)

Kaye, G.R., *The Astronomical Observatories of Jai Singh* (Archaeological Survey of India, New Imperial Series, Vol. XL, Calcutta, 1918)

King, David A., *Astronomy in the Service of Islam* (Variorum, Aldershot, 1993)

Koestler, Arthur, *The Sleepwalkers* (Penguin, London, 1964)

Lippincott, Kristen, *The Story of Time* (National Maritime Museum, London, 2000)

Lloyd, Steven A., *Ivory Diptych Sundials 1570–1750* (Harvard University, 1992)

MacGibbon, David, and Ross, Thomas, *The Castellated and Domestic Architecture of Scotland from the Twelfth to the Eighteenth Century* (David Douglas, Edinburgh, 1887)

Mancinelli, Fabrizio, *La Torre di Venti in Vatican* (Libreria Editrice Vaticana, 1980)

Mills, Allan, 'Isaac Newton's Sundials', *Antiquarian Horology* XX (Summer 1992)

Miniati, Mara, *Museo di Storia della Scienza* (Giunti, Florence, 1991)

Morton, Alan Q., *Science in the Eighteenth Century* (Science Museum, London, 1993)

Needham, Joseph, *Science and Civilisation in China: Mathematics and the Sciences of the Heavens and the Earth* (Cambridge University, 1959)

Neppi, Lionello, *Palazzo Spada* (Editalia, Rome, 1975)

Neugebauer, O., *The Exact Sciences in Antiquity* (Brown University, 1957)

O'Kelly, Michael, *Newgrange* (Thames and Hudson, London, 1982)

Pantanali, Aurelio, el al., *Meridiane del Friuli-Venezia Giulia* (Forum, 1998)

Peyrefitte, Alain, *The Collision of Two Civilisations: The British Expedition to China in 1792–4* (Harvill, London, 1993)

Richards, E.G., *Mapping Time: The Calendar and its History* (Oxford University, 1998)

★ Rohr, René, *Les Cadrans Solaires* (Editions Oberlin, 1986)

★ _____, *Sundials: History, Theory and Practice* (University of Toronto, 1970)

Sawyer, Frederick (ed.), *The Analemmatic Sundial Source Book* (North American Sundial Society, 2004)

Scarr, M.M., *The Dial in Old Court, Queens' College* (Queens' College, Cambridge, 1988)

Sedilot, J.-J., and Sedilot, L.-A., *Traité des Instruments Astronomiques des Arabes* (republished by Johann Wolfgang Goethe University, 1988)

Somerville, Andrew, *The Ancient Sundials of Scotland* (Rogers Turner, Edinburgh, 1990)

Souden, David, *Stonehenge* (Collins and Brown, London, 1997)

Stanier, Margaret, *Oxford Sundials* (Somerville College, Oxford)

Sundial Glossary (British Sundial Society, 2004)

Turner, Gerard l'E., *Elizabethan Instrument Makers: The Origins of the Trade in Precision Instrument Making* (Oxford University, 2000)

Volwahsen, Andreas, *Cosmic Architecture in India* (Prestel, Munich, 2001)

Watkin, David, *Athenian Stuart* (George Allen & Unwin, London, 1982)

★ Waugh, Albert, *Sundials: Their Theory and Construction* (Dover, London, 1973)

The entries marked ★ are recommended for the construction of sundials.

GLOSSARY

Some of the following definitions contain simplifications.

analemma The figure-of-eight curve which is sometimes drawn round or in place of the noon line to signify the precise moment of noon clock time. Confusingly, the word is also used with a different meaning to describe the analemmatic dial (qv).

analemmatic dial A dial that has a gnomon movable against a date scale and whose shadow is cast on to time points marked on an elliptical ring. Sometimes these are constructed in public places where a standing human can act as the gnomon.

armillary dial A dial based on an armillary sphere (qv).

armillary sphere A skeleton model of the celestial sphere (qv), first described by the Alexandrian astronomer Claudius Ptolemy. It usually has rings representing the equator, the tropics, the ecliptic and Arctic and Antarctic circles.

astronomical ring dial The dial developed from the armillary sphere by Gemma Frisius in 1557.

azimuth The angular direction of the sun, usually measured in degrees.

Babylonian hours Used for measurement from one sunrise to the next, with an equal hour system.

canonical hours The seven times of day used to define the services of prayer, which had been developed by St Benedict in the sixth century. They were a form of unequal hours and survived for use in the Christian church for eight hundred years.

celestial sphere The dome of the heavens imagined by the ancients as a crystal sphere above us.

Chaucer's cylinder see shepherd's dial

conical dial A classical dial in which the time and date information is marked on the inside of part of a cone. The invention is attributed to Dionysodorus (*c*.250–*c*.190 bc).

crescent dial A modification of the universal equatorial dial (qv). It has two crescent-shaped gnomons instead of an aperture.

cylindrical dial see shepherd's dial

declination (1) The angle of the sun measured above or below the celestial equator. At the equinoxes declination is zero, at the solstices approximately plus or minus 23.45°.

declination (2) The angle a vertical dial makes with south. Walls declining west of south have a positive value, those declining east a negative value.

declining dial A vertical dial tilted towards west or east

diptych dial A portable dial with two faces hinged together, and with a gnomon in the form of a cord running between them.

displacement Few dials are located precisely on a time meridian. For each degree west of a time meridian that a dial is displaced, 4 minutes must be added to the dial reading. For dials to the east, the value must be subtracted.

ecliptic The plane which the earth's orbit traces round the sun during a year. The orbits of the moon and planets are close to this plane, and so eclipses take place near it.

equal hours The time system in universal use today where each hour of day and night is equal in length.

Equation of Time (EoT) The difference in any place between sun time (apparent solar time) and mean solar time.

equatorial dial A dial in which the dial plate or circle is set parallel with the equator and the polar pointing gnomon is perpendicular to it. The finest example is the Henry Mo0re sculpture in the frontispiece. Note that many authorities insist that this dial should be called equinoctial, but the use of this word makes the dial harder to understand, and the term equatorial has been used throughout this book.

equinoctial dial see equatorial dial.

equinoxes Those days (20 or 21 March, the vernal or spring equinox, and 22 or 23 September, the autumn or fall equinox) when the lengths of day and night are equal and the sun rises due east and sets due west all over the earth.

great circle A circle on the surface of the earth whose diameter is the same as the earth.

Greenwich Mean Time (GMT) The basis for civil time in the UK, based on the Greenwich meridian. Other countries use different time zones. The USA has six – Alaska, Pacific Standard, Mountain Central, Standard, Eastern Standard and Atlantic.

Gregorian calendar The calendar introduced by Pope Gregory XIII in 1582, which today is universally used throughout the world. He continued the use of a leap year every four years, but added a new rule that there should not be a leap year in those century years that are not divisible by 400 (e.g. 1700). These two simple rules produce such accuracy that they will lead to the calendar running ahead of the seasons by only one day in the course of the next 3,000 years.

hemicyclium Very similar to the hemispherium but with the unnecessary part of the dial face cut away. Vitruvius attributes the invention of this dial to Berosus, and the hemispherium to Aristarchus, but some believe it was the other way round.

hemispherium A classical dial on which the time lines were engraved inside a bowl or hemisphere. Often attributed to Berosus, the Chaldean astronomer who lived in the third century BC.

horizontal dial A dial created on a horizontal surface.

hour angle (HA) The angle corresponding to the sun's position in its daily orbit. Each hour is equal to 15 degrees.

Italian hours Used for measurement from one sunset to the next, with an equal hour system.

Julian calendar The calendar introduced by Julius Caesar in about 45 BC at the behest of Sosigenes. He decreed the year to be 365 days long and 366 days every fourth (or leap) year.

Local Apparent Time (LAT) The measure of time at any location based only on the movement of the sun. Referred to in this book as sun time or local solar time.

Local Mean Time (LMT) Solar time which is corrected for the Equation of Time (EoT) but not for the displacement of the observer from his time meridian (qv).

longitude The location on the earth of a place east or west of the Greenwich meridian. Solar longitude is the position of the sun during the year measured along the ecliptic.

meridian The line of longitude that passes through the observer's location or its representation on the dial. Time meridian means the line of longitude which defines the time zone of any area.

nodus A point on a gnomon, often a small sphere, or a hole in a disc, which indicates date on a dial face.

noon line The line on a dial signifying the hour of noon.

pillar dial see shepherd's dial. The word is also used to describe monumental dials on tall pillars.

plane dial A dial on any flat surface.

planetary hours see unequal hours

Platonic solids The five regular solids, or those which have their faces, sides and angles equal. They are the cube, tetrahedron, octahedron, dodecahedron and icosahedron. There are four other regular solids, which were not known to the Greeks.

Pole Star The star which is close to the north celestial pole, and round which the other stars appear to revolve at night.

polyhedral dial. A dial with many faces, sometimes horizontal, scaphe, vertical, declining and reclining (qv).

Precession of the Equinoxes The long term wobble in the direction of the earth's axis, which Newton explained was caused by the gravitational tug of the moon.

reclining declining dial A dial with both reclining and declining features.

reclining dial A south-facing vertical dial tilted towards north or south.

reflecting dial A dial which operates from a reflected spot of light. Sometimes known as a catoptric dial, from the Greek for mirror.

ring dial An altitude dial in the form of a ring which is suspended and turned so that a small aperture faces the sun.

scaphe The classical Greek word for a boat, used to describe a dial in the form of a shallow bowl.

seasonal hours see unequal hours

shepherd's dial A pillar dial which measures time from the altitude of the sun.

solstice The instant, in midsummer or midwinter, when the earth's axis is tilted to its most extreme angle in relation to the sun.

style The actual line in space, only part of a gnomon, which generates the shadow edge to indicate time.

temporal hours see unequal hours

trigon An instrument for setting out the date curves on a plane dial.

tropic of Cancer The northern tropic which represents for people living in the northern hemisphere the extreme of the region where the sun can reach the zenith (qv) at midday.

tropic of Capricorn The southern tropic – opposite of Cancer.

unequal hours The time system in which each period of daylight is divided into hours of equal length, same for nighttime. It follows that a summer daylight hour is longer than a winter one for people who live north of the equator.

universal equatorial dial A portable self-orienting equatorial dial with an aperture gnomon, and movable suspension point to adjust it for latitude. Also known as a universal equinoctial dial.

vertical dial A plane dial on a vertical surface.

zenith The point on the celestial sphere vertically above the observer.

zodiac The imaginary band in the night sky in which the moon, planets and the twelve constellations are always located. The sun is also located in this band.

ACKNOWLEDGMENTS

The Publishers have made every effort to contact holders of copyright works. Any copyright holders we have been unable to reach are invited to contact the Publishers so that a full acknowledgment may be given in subsequent editions. For permission to reproduce the images on the following pages, and for supplying photographs and artworks, the Publishers thank those listed below.

Archivio Segreto Vaticano: 90–1; **Paolo Alberi-Auber** (www.ingauber-meridiane.it): 33, 121 right, 125 right; **Riccardo Anselmi** (riccardoanselmi@libero.it): 21; **Douglas Bateman**: 62 right (courtesy of QinetiQ); **Bayerisches National Museum**: 70; **Dr Edwin V. Bell/NASA/ NSSDC** (ed.bell@gsfc.nasa.gov): 56 above left; **Bibliothèque royale de Belgique**: 49 right; **www.bridgeman.co.uk**: 12 above, 22, 39 below, 67, 77 above right (Louvre, Paris), 78 (National Gallery, London), 87 (Archivo del Stato, Siena), 112 (Derby Museum and Art Gallery); **bpk/Staatliche Museen zu Berlin-Aegyptisches Museum (photo Margarete Buesing)**: 15; **By permission of the British Library**: 24, 37 left, 94 right–95; **© The Trustees of the British Museum**: 19 above right (GR 1821.3-1.1), 72 above (MME 1896, 12-14, 1); **Michael Burke** (www.burkephotography.biz): 122 below; **Martial Casanova**: 124 below; **Christie's Images**: 49 left, 50–1, 54, 55 above, 71 above left, 72 below, 103 below right, 116; **Cornell University News Service**: 127; **Mike Cowham**: 106 above; **Christopher St J.H. Daniel** (www.sundialdesign.co.uk): 62 left, 63; **John Davis** (www.flowton-dials.co.uk): 114 right–115; **Department of the Environment, Heritage and Local Government, Dublin** (www.environ.ie): 12 below; **David Harber** (www.davidharbersundials.com): 16, 28; **Claude Hartman**: 84 below; **Min Hogg**: 46 left; **Hendrik Hollander** (www.analemma.biz): 79 above; **Instituto e Museo di Storia della Scienza, Florence** (www.imss.fi.it): 71 below, 71 above right, 77 below; **Thomas Jefferson Foundation** (www.monticello.org): 74 above; **Ben Jones** (www.benjonessundials.co.uk): 64, 126; **Gerald Laing** (www.geraldlaing.com) **and Ken Mackay**: 120 left; **Michèle Lapointe and René Rioux** (www.aei.ca/~mlapointe2): 79 below; **© Andrew Lawson**: 120 right; **www.longwoodgardens.org**: 108 below; **Mark Lennox-Boyd**: 10, 17 above left, 18, 19 above left, 19 below right, 20 right, 27 above, 31, 41 right, 45, 65, 68, 81, 94, 98 above, 99 above, 103 below left, 109, 128, 129, 130, 131, 133, 135; **Jonathan Lynch** (www.lynchphotography.co.uk): 2, 16–17 below, 17 above right, 27 below, 55 below, 111; **Frans Maes** (www.fransmaes.nl/sundials): 48; **Joanna Migdal** (www.sundialsomething.com): 57, 58, 101 below right; **Allan Mills** (am41@leicester.ac.uk): 19 below left; **Reproduced by permission of the Henry Moore Foundation**: 56 centre right, 56 below right; **James Mortimer**: 74 below; **Tony Moss** (www.lindisun.demon.co.uk): 114; **Museum of the History of Science, Oxford** (www.mhs.ox.ac.uk): 77 above left; **Quentin Newark, Atelier Works** (www.atelierworks.co.uk) **and Frank King** (fhkl@cam.ac.uk): 108 above right; **Reproduced by kind permission of His Grace the Duke of Norfolk, Arundel Castle**: 104 above; **© Oxfordshire County Council Photographic Archive – Thomas Photos**: 104 below right; **Courtesy of Trevor Philip & Son Ltd, London**: 93, 103 above, 104 below left, 110; **© Eve O'Kelly**: 13; **Kate Pond** (www.vermontsculpture.com): 123; **Private collection**: 6–7 (background), 9, 29, 32, 36, 37 right, 39 above, 40, 41 left, 43, 44, 46 right, 47, 52–3, 56 above right, 60, 73 left, 73 right, 84 above, 97, 98 below, 101 above, 101 below left, 108 above left, 117, 119, 121 left; **Helm Roberts** (www.helmr.com): 125 left; **Royal Astronomical Society** (www.ras.org.uk): 66, 75, 102; **Colin Russell**: 106 below; **Denis Savoie**: 124 above; **© Tate, London 2006**: 107, 121 centre; **Turtle Bay Exploration Park** (www.turtlebay.org): 122 above; **Phil Wherry** (www.wherry.com/photos): 5; **Luke White** (www.lukewhite.com): 80, 82, 83; **Wilson Publications and Associates Ltd** (www.wilson-publications.co.uk): 25; **Ono Yukio**: 20 left, 35, 85, 99 below

Many of the above gave freely of their photographs and I am most grateful to them, but others also helped in other ways. In particular I must repeat my thanks to Christopher St J.H. Daniel and Mario Catamo. I would also like to thank for different varieties of help: Paolo Alberi Auber, Anthony Arfwedson, Tony Belk, Selina Hastings, Peter Hingley, Min Hogg, Nick Hornby, Mohammed al-Khouly, Frank King, Patricia Lennox-Boyd, Frans Maes, James Mitchell, Colvin Randall, Margaret Stanier, Zahed Tajeddin and Adam Williams. Finally, my very great thanks to the staff at Frances Lincoln, and in particular to Caroline Clark for her design, which required her considerable ingenuity to ensure that the illustrations were inserted close to the relevant text, and to my editor Michael Brunström for his support and the many improvements he made.

Mark Lennox-Boyd

INDEX

Page numbers in *italic* refer to the captions to the illustrations. Page numbers followed by *g* refer to glossary definitions.

A

Accademia Olimpica 136
accuracy of sundials 59, 60
Age of Enlightenment 111–13
alchemy 29, 105
All Souls College, Oxford 105, *105*
altitude dials *72, 85*
The Ambassadors 76, *77, 78*, 123
analemma 61, *62*, 138*g*
analemmatic dials 108–111, *108–9, 110,*
   138*g*
Andronicus of Cyrrhus 23
anemoscopes 89, *91*
angle of style 128
Anselmi, Riccardo *20*
Antiquities of Athens 23, *24*
Antistius Euporus, M. 32, *33*
Apparent Solar Time 59–60, 89
Aquileia, Italy 32, *33*
Arab astronomers 38, 42
Arabic words 38, 42
Aristarchus of Samos 15, 20, *25,* 26
armillary dials 26, *29,* 138*g*
armillary spheres 26, *27, 59, 117,* 138*g*
Arroyo Grande, California *85*
Arundel, Thomas Howard, 2nd Earl
   102, 103, *103*
As You Like It 72
astrology 29–30
   see also zodiac
astronomer priests 88, 89, 92, 100, *100*
astronomical ring dials 26, *19,* 138*g*
astronomical tables 134
Augsberg 51, *52, 72, 77, 105*
Augustus 34, *126*
azimuth 134, 138*g*
azimuth dials *114*

B

Babylonian astronomers 14–15, 30
Babylonian hours 22, 42, *43, 44,* 138*g*
Bacon, Roger 93
Bartholomew, Abbot of Aldersbach *72*

Bateman, Doug 61, *62*
Beano, Italy *121*
Beata Beatrix 121, 125
Bedos de Celles, François *108*
Beer Glass Dial *76, 79*
Beijing observatory 100, *100*
Benares, India *97*
Berosus 14–15
Bewcastle, Cumbria 38, *41*
Bianchini, Francesco 92, 93, *94*
Big Horn Medicine Wheel, Wyoming 10
Bion, N. *55*
Bizard Island Municipal Library, Quebec, *79*
Blake, William *107*
Blateyron, François 136
Bonfa, Father 73, *74, 76*
Boyle, Charles, Earl of Cork and Orrery *113*
Brahe, Tycho 66, *67,* 68
British Museum *18, 72*
British Summer Time 60
British Sundial Society (BSS) 121, 132, 136
Bronze Age rondels 10
Brou, France 111
Brunelleschi, Filippo 73
Buchner, Edmund 36, *37*
Buonsignori, Stefano *70*
Butterfield, Michael *55*

C

Calatrava, Santiago 121
calendars
   Gregorian 88–9, 139*g*
   Islamic 42, *46*
   Jewish 87
   Julian 86, 88, 139*g*
   Mayan 11
camera obscura *47, 81,* 93
Campus Martius, Rome 34, 36, *36–7*
Cancer, tropic of 26, 140*g*
cannon dials *93*
canonical hours 38, *41,* 45, 138*g*
   see also unequal hours
Capricorn, tropic of 26, *53,* 140*g*
cardinal points 8–9, *10*
carpe diem 120–21
celestial sphere *27,* 138*g*
chalice dials *72,* 73
Charles I *72*
Chartres cathedral 38, *41*
Chaucer, Geoffrey 59, 76

Chaucer's cylinder *see* shepherd's dials
China 10, 100
Christianity, and time measurement
   38–42, 86–93
Christoph, Hans 51
chronometers 126
Church *see* Christianity
Circignani, Niccolò 89, *91*
Clavius, Christopher *87,* 88, 89, 100
Cleanthes 20
Clement XI, Pope 92, 93
Cleopatra 34, 86
Clifford, Lady Anne 121, *121,* 124
clocks 51, 59, *92, 116, 120,* 126
Codrington Library, Oxford 105
Cole, Humphrey 54
Colsterworth church 107, *107*
Commission for Calendar Reform *87,* 88
compasses
   in hand-held dials 51, *51, 53, 54, 55*
   inaccuracies in 102, *110,* 134
   sun *110*
   see also cardinal points
compendia 51, *52–3, 54*
computer programs 136
conical dials 20, *20,* 138*g*
Constantine I 86–7
constellations 23, 30
   see also zodiac
Copernicus 100, 112
crescent dials *105,* 138*g*
cross dials *73*
cylindrical dials *see* shepherd's dials

D

Daniel, Christopher St J.H. *62*
Daniell, Thomas and William *97*
Danti, Egnatio 89
date 76, 86
   calculations for sundials 132–4
   Easter 43, 86–8, 89, 92–3
   see also zodiac
Davis, John *114*
daytime measurement 14, 22, 38
de Vaulezard 108
declination *30,* 132, 134, 138*g*
declining dials 130, 134, 138*g*
Delhi 97, *97, 99*
design of sundials 42, 73, 120, 121, 126
   calculations 128–36